Springer-AAS Acoustics Series

Series Editor

Dr. Marion Burgess, School of Engineering and Information Technology, University of New South Wales, Sydney, Australia

This series publishes peer reviewed high quality monographs and contributed volumes on all topics in acoustics. Books in this series range from those focused on a particular aspect of acoustics and vibration to practical handbooks covering a range of topics. The advantage for authors is that the inclusion of a book as part of this series will demonstrate high quality content of the book. While this series encourages authors and topics that are relevant to the Australasian region, proposals for contributions to the series are not be restricted to this region.

Irene van Kamp · Fred Woudenberg

Editors

A Sound Approach to Noise and Health

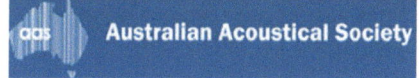

Editors
Irene van Kamp
National Institute for Public Health
and the Environment
Bilthoven, Utrecht
The Netherlands

Fred Woudenberg
Municipal Health Service of Amsterdam
Amsterdam, Noord-Holland
The Netherlands

ISSN 2948-2062   ISSN 2948-2070   (electronic)
Springer-AAS Acoustics Series
ISBN 978-981-97-6120-3   ISBN 978-981-97-6121-0   (eBook)
https://doi.org/10.1007/978-981-97-6121-0

This Springer imprint is published by the registered company Springer Nature Singapore Pte Ltd.
The registered company address is: 152 Beach Road, #21-01/04 Gateway East, Singapore 189721, Singapore

If disposing of this product, please recycle the paper.

Preface

It was back in 2018 that we discussed and later signed a contract with Springer Nature to edit a book of approximately 200 pages on health and noise. We were made aware of the Acoustics series at the Inter-noise conference in Crete by Loy d'Silva and Marion Burgess. Preparations were made and authors were approached. Originally the book was planned to be published in 2020, but then COVID came. A new plan has been made and now after 6 years the book is there. And maybe typical for the field, but still surprising, the topic is not outdated.

Sound plays a key role in human life, in our orientation in time and space, a sense of safety or threat, survival, but also in the expression of emotion, in poetry, and even in codetermining the meaning of life.

The role of sound in our daily life is highly underestimated or just taken for granted. So are the beneficial and harmful effects. Our bodies respond to sound, according to some in response to the meaning we give them, whilst others would argue that the responses are there whether the sounds are wanted or unwanted. This is one of the reasons to adapt the definition of noise. The original definition of noise being unwanted sound was recently extended with or/and harmful sound: some effects happen whether we like the sounds or not. At the same time sounds obtain meaning in a certain physical and social and personal context and in that respect the field differs very much from other environmental hazards such as water and air pollution.

Except in extreme cases where loud noises can damage our ears, unwanted sounds are harmful to health by repetition (Dripping water hollows out stone, not through force but through persistence). And to further complicate the story: soundscapes are dynamic and change in time and so do their meanings. Soundscapes will inevitably change in the future.

It is not an easy task to account for all these aspects in studying environmental sound, but also to govern them. Most studies and policies are aimed at single sources and single effects and economic considerations play an important role. The soundscape approach is a response to this, but has not focused on the larger scale beneficial and harmful effects of chronic exposure to unwanted or harmful sounds. However, it has stimulated to move towards an interdisciplinary approach, linking urban planning to the field of sustainability and public health, paying attention to harmful and beneficial effects and becoming aware of the sonic environment as a first step in improving them.

Having worked in this field for almost 30 and 40 years, including collaborating on several books in the Netherlands in the past, we as the editors of this book felt a strong need to give the whole story in one place, in the hope that the different disciplines would meet. Our goal is to share with a broader audience the many aspects of this fascinating field of sound and make them aware that we all contribute, professionally and personally. We hope our 'A Sound Approach to Noise and Health' will help create a future in which much less people suffer the negative effects of noise and many more people enjoy the sounds that restore their well-being in a space defined by sound, and sound surrounded by silence.

Bilthoven, The Netherlands   Irene van Kamp
Amsterdam, The Netherlands   Fred Woudenberg

Acknowledgements

We gratefully thank Marion Burgess for drawing our attention to the acoustics series of Springer Nature and reviewing all chapters. Lex Brown for his critical comments to earlier drafts of the outline and introductory chapters, and Hans van Leeuwen for reviewing the entire book. Guus de Hollander, we thank you for your critical notes on health impact assessment. We also greatly appreciate the collaboration with 'Springer Nature' in particular Loy d'Silva and Vinothini Elango.

This open-access book was financially supported by:

- The National Institute for Public Health and the Environment, Ministry of Health, Welfare and Sport in the Netherlands, BDR C&D/DIV;
- Expertise Centre Noise, Centre for Sustainability, Environment and Health RIVM Ministry of Infrastructure and Water Management; the Netherlands;
- The UK Health Security Agency formerly Public Health UK, Noise and Public Health;
- Municipal Health Service Amsterdam, Netherlands;
- BAFU, Swiss Federal Office for the Environment (FOEN) Noise/NIR Division.

Contents

Chapter 1
Introduction

Fred Woudenberg and Irene van Kamp

Abstract A sound approach to noise and health is interdisciplinary. Next to the negative health effects of unwanted sound that historically get most attention, it should include the beneficial effects of sound and insights from the physical, medical, psychological, economic, spatial planning, governance and art disciplines. This chapter puts the issue of noise and health in a broader societal context and gives an outline of the present book.

> Nature is for the idle or contented.
>
> And then: what's to be found here, nature wise?
>
> A scrap, perhaps, of woodland, post-stamp size,
>
> A hillock with some cottages against it.
>
> Give me the unrelieved grey city roads,
>
> The waterside imprisoned into quays,
>
> The clouds, at their most beautiful always
>
> As, window-framed, along the sky they float.
>
> Everything's much when much is not expected.
>
> Life hides its miracles till, without warning,
>
> They are unfolded to be marveled at.
>
> Along these lines I silently reflected,
>
> Bedraggled, on a drab and drizzly morning,
>
> Quite simply happy, in the Dapperstraat.
>
> (JC Bloem 1946)

F. Woudenberg
Municipal Health Service, Amsterdam, The Netherlands
e-mail: fwoudenberg@ggd.amsterdam.nl

I. van Kamp (✉)
National Institute for Public Health and the Environment, Bilthoven, The Netherlands
e-mail: Irene.van.kamp@rivm.nl

Fig. 1.1 Photo of a house in the Dapperstraat in Amsterdam with the first lines of the poem by Bloem in many languages. Photo by Edwin van Eis, Municipality of Amsterdam, permission granted by the photographer, Amsterdam 2024

This translation of a famous Dutch poem published in 1946 is about the (melancholic) state of happiness of a person living in the inner city of Amsterdam at the time. It is a more profound state than the happiness of somebody enjoying nature or what is left of it in an already then densely populated part of the Netherlands. In this poem, Bloem expresses his love for the city more than anything else (Fig. 1.1).

The poem beautifully describes the utterly subjective relation between a person and his immediate environment. This is an all-time relationship also applicable to current times and housing. The place in which we live matters and the focus on the residential situation of current environmental policies is fully justified. On average, people spend some 16 hours per day at home and during the pandemic the time spent at home has gone up considerably and might remain high. For most people, a quiet home is crucial and considered the most important place to relax and restore from daily pressures. In his groundbreaking work "livable streets" Appleyard (1980) draws attention to the seemingly trivial fact "that nearly everyone in the world lives on a street". People have always lived on streets. As Appleyard writes:

> They have been the places where children first learned about the world, where neighbors met, the social centers of towns and cities, the rallying points for revolts, the scenes of repression. But they have also been the channels for transportation and access; noisy with the clatter of horses' hooves and the shouts of their drivers, putrid with dung, garbage and mud, the places where strangers intruded, and criminals lurked.

At that time, and also nowadays, the harmful effects of traffic pose one of the most widespread and threatening problems according to Appleyard, even more so than crime. A multitude of interconnected physical and social features, settings and personal aspects such as mood and activity determine how people feel in a certain place at a given moment. Many people do not want to leave the place where they grew up, while others cannot wait to get away from it, illustrating the subjectivity of how people feel in a certain place. Despite this common attachment to the place of birth, millions of people leave the place of their cradle for a whole range of reasons. The bulk of people end up in a city, but in certain countries, a reverse trend can be observed as well with people moving from urban to rural areas. The post-pandemic trend towards online working might cause yet another trend as it offers the opportunity to work from anywhere. Remote working has even led to a breed of digital nomads with some countries providing special visas for them (https://www.byevisa.com/tomorrows-top-remote-working-destinations/).

At this point in time, approximately 57% of the world's population lives in urban areas, and this percentage is expected to increase to 68% by 2050. In 2007, for the first time in human history, more people globally lived in urban than in rural areas (https://ourworldindata.org/urbanization#:~:text=More%20than%204%20billion%20people%20live%20in%20urban,world%20lived%20in%20urban%20than%20in%20rural%20areas). Not so long ago, in the early 50s, only 30% of the world's population was living in urban areas. In many parts of the world, the city offers economic opportunities that are not to be found elsewhere and offers amenities such as medical care, culture, amusement and the opportunity to meet people (in one word: density of interaction). But the city can also be a nasty place to live in with its crowds, its lack of social cohesion, lack of green space and tranquility, bad air and the often neglected omnipresent unwanted sound. These unwanted and wanted ones are the focus of this book.

1.1 Sound Is Crucial for Life but Highly Underestimated

Our nature-averse poet Bloem (1946) might be an exception to the rule. There is increasing evidence supporting the so-called biophilia (love of nature) hypothesis stating that people are innately attracted to environments in which the human species evolved: green spaces in which they felt connected to all other life forms (Wilson

1986). Whether this experience of the environment has to be real or could also be virtual is another question: it has been shown that a restoring effect can also be achieved when people simply see pictures or movies of nature, hear bird sounds in airport lounges or enjoy a work of art (Grinde and Patil 2009).

Also, the appreciation of nature is not static and universal. Our innate attraction is modulated by our experiences and beliefs. In medieval times, nature was considered threatening, home to dangerous animals and aggressive humans. The same holds for the sounds of nature. Environmental historian Coates (2005) points out that what we think of as noise is as much a matter of ideology as of decibels. Some people have long been annoyed by unwanted sounds, usually referred to as noise, while for others the sounds of human activities are signs of progress and prosperity and security. Coates (2005) describes how European settlers in America evaluated the noise of an axe striking a tree as an "aural victory over howling wilderness". He also describes how to nineteenth-century modernists, mechanical sounds and the noisy bustle of commerce bespoke prosperity while quiet "was synonymous with indolence, backwardness, and stagnation". For many, the sounds of the great reconstruction after World War II and the rapid and ubiquitous industrialization sounded like music, bringing prosperity and security. Currently and maybe connected to the increased affluence in many countries which many of its inhabitants seem to take for granted, people are getting less tolerant of ambient sounds, be it mechanical or human. The most plausible explanation for this trend might be found in comfort expectations and a lifestyle of health and sustainability in combination with a growing awareness of the health risks of noise (Wunderli and Brink 2022).

1.2 Sound and Nature

It is not fully clear which factors are responsible for the above-described intrinsic beneficial effects of nature. There are strong indications that visual aspects are key; people having undergone surgery, for example, recover faster if they have a view on green from their hospital bed. Even a view on wallpaper of a forest seems to work and sounds of nature can have similar beneficial effects (Ulrich 2008). It explains why most people have a preference for natural sounds (Brown 2010) like the rustling of leaves and the singing of birds, the sounds of rippling water and why they often detest mechanical sounds. In this respect, our prehistoric ancestors lived in paradise. Noise as unwanted sound was back then probably "limited" to sounds signaling danger, the roaring of a saber tooth tiger, screams of fear from tribe members—in specific children—or war cries from an attacking nearby tribe. Nowadays, mechanical sounds which may be perceived as threatening or at least mask the pleasant sounds signaling safety, are omnipresent in the modern urban jungle. Transport noise and in particular air traffic noise is the main reason why it is hard to find places where mechanical sound cannot be heard for a prolonged time. Acoustic ecologist Gordon Hempton (Hempton and Grossmann 2009) has been looking for places without mechanical noise for decades and drew to the conclusion that silence is an endangered species on the verge of extinction.

1.3 Since When Does Noise Exist?

As indicated above noise is unwanted (and/or harmful) sound. It is primarily related to sound levels and for many people loud sounds are unpleasant and thus experienced as noise. But loud sounds can also be very pleasant: just ask the people who pay a lot of money to attend a concert of Metallica. Whether sound is considered as noise depends to a large extent on the receiver/perceiver. The experiences of the receiver are partly innate and to a large extent shaped by family, social ties and culture. This leads to the question if cultures differ in their experience of sound and if people have always experienced noise or whether it is to be considered as a modern invention.

Probably humans have always been bothered by certain sounds. Many animals are bothered too. This is well described for whales, who for their social life depend on long-range vocal communication (https://www.nhm.ac.uk/discover/news/2022/july/underwater-noise-pollution-risking-lives-whales-dolphins.html). It is known that birds adapt their communication to ambient noise levels (Hu and Goncalo 2009) and that animals like the urban black-tufted marmosets living in noisy urban areas select their homes based primarily on ambient noise levels (Duarte et al. 2011).

1.4 Noise and Technology

The history of noise is closely intertwined with the development of technology. For millennia important sources of noise were the human voice and the sound of wooden and iron wheels. It was the sound of wooden wheels riding over cobblestone streets that led to complaints and made life in ancient Rome for some acoustically unbearable. Carts were pulled by oxen, horses and other animals and their hoofs added to the noise. Documented proof of earlier noise annoyance is of course hard to find. There are no records before the origins of writing in the 4th millennium BCE.

Archaeoacoustic studies show that our prehistoric ancestors were well aware of acoustics. As Steven Waller writes about our ancestors (Waller 2019):

> He selectively chose echoing environments to produce cave paintings and canyon petro-glyphs. The images they produced were descriptions of echo spirits that figured in echo myths. The cultural perception of echoes was that these "extra" sounds were considered mysterious answers from sacred beings worthy of worship.

Sound or noise are not mentioned in the famous code of Hammurabi, a Babylonian legal text composed c. 1755–1750 BC. It contains rules on a large number of crimes and offences. It was preceded by the Mesopotamian code of Ur-Nammu, the oldest known law code surviving today, written in Sumerian somewhere between 2100 and

2050 BCE. No reference to noise is made in that either. This is remarkable since these old rules mainly concerned violence, sexual matters and housing property, all potentially accompanied with sound.

Reference to noise is made in 'Atrahasis', a Mesopotamian epic of the second millennia BC (Simona et al. 2021). In the epic, Enlil, a Mesopotamian man-like god wants to eliminate humankind due to the noise disturbance caused by people, which deprives him of sleep. From neo-Babylonic times, several hundred years BC, dates a story (https://www.asor.org/anetoday/2015/08/policemen-in-1st-millennium-bc-bab ylonia/) about a policeman in the city of Uruk, a certain Balātu, who arrested a group of persons in front of a tavern because of a nighttime noise violation. The first known regulation of noise stems from the sixth century before Christ (https://mikegoldsmith. weebly.com/history-of-noise.html). The council of the province of Sybaris, a Greek colony in the Aegean, ruled that potters, tinsmiths, and other tradesmen must live outside the city walls because of the noise they make. Also, roosters had to be taken out of the city premises. The example of Sybaris shows that occupational noise was a main source of annoyance as well. Brian Hunt (https://www.aeraweb.org/blog/2009-field-season/the-sounds-of-antiquity/), visiting the Giza pyramid, pondered about the noise the workers had to face during construction in the twenty-sixth century BC. In the fifth century BC Hippocrates was possibly the first to clearly identify tinnitus and link it to noise (https://mikegoldsmith.weebly.com/history-of-noise.html). In medieval times in Europe, the sound of church and other bells were added to these occupational sounds.

Ancient Rome was notorious for its noise and smell. Julius Caesar ruled that "no one shall drive a wagon along the streets of Rome or along those streets in the suburbs where there is continuous housing after sunrise or before the tenth hour of the night" (https://mikegoldsmith.weebly.com/history-of-noise.html). With this rule, Caesar seemed to worry about daytime noise rather than nighttime noise and sleep disturbance as we do nowadays. Because Rome was so crowded, carts with produce and products were only allowed into the city after dark. The noise at night in Rome could be deafening. A Roman satirist, Juvenal, wrote about the impossibility of sleep in ancient Rome, except for the wealthy (http://www.antiquitatem.com/en/ancient-cities-and-noise-deafness-sybari/):

> For what sleep is possible in a lodging? Who but the wealthy get sleep in Rome? There lies the root of the disorder. The crossing of wagons in the narrow winding streets, the slanging of drovers when brought to a stand, would make sleep impossible for a Drusus or a sea-calf.

One of the most remarkable noise regulations, at least from a contemporary perspective, was issued in noisy London at the end of the sixteenth century. It was for men not allowed to beat their wives after 9 PM as described in the bye-law forbidding any "suddaine out-cry... in the still of the Night, as making any affray, or beating his Wife, or servant, or singing, or reveling in his house, to the Disturbance of his neighbours" (https://mikegoldsmith.weebly.com/history-of-noise.html).

1.5 Future Noise

The noise environment or soundscape changes continuously. Electrification of traffic will change road noise, although not everybody expects lower sound levels. At 50 km/h electric cars produce more tire noise because they are on average heavier than fossil-fueled cars. Electromobility helps reducing noise, but only at lower speeds (30 km/h) and/or with lighter vehicles. At the same time and due to these developments, the lower frequencies (LFN) become more salient/noticeable. New sources of noise like wind turbines and heat pumps are emerging due to the energy transition and climate change oriented measures (Kamp 2022). Climate change leads to architectural adaptations in urban areas, increased use of ventilation/cooling systems and behavioral changes that may affect noise levels like going outdoors more often or staying up later. Most of these are part of the energy transition and could affect noise exposure and related perceptions and health/well-being. A review written by the Netherlands Health Council (NHC) (Netherlands Health Council 2020) confirmed and partly expanded the importance of earlier detected noise issues related to energy transitions. Noise issues that came forward are the increased use and complexity of electronic appliances such as collective heating systems, heat pumps, and cooling systems. These sources can lead to higher sound levels or have specific, often low-frequency annoying components affecting well-being and health.

1.6 Decibels Versus Perception

In most noise and health studies environmental noise is considered a pollutant, a somewhat unavoidable waste product and a stressor which can lead to negative responses. In current European noise policies regulations are tied to noise levels expressed in decibels and annual averaged day evening night levels L_{den} and L_{night} are used to predict the health effects. This stimulus-response approach is at the base of most noise research and policy in which the emphasis lies on threshold levels, standards and interventions aimed at reducing levels. However, this almost exclusive attention on physical noise metrics is starting to shift somewhat towards attempting to understand the role of context in people's perceptions of, and reaction to, acoustic environments. The so-called soundscape approach considers the acoustical environment more broadly as a resource, not merely as a waste product (see Table 1.1). It shifts the focus from the physically measured levels of exposure towards considering more broadly the perception of sound—by making a distinction between wanted and unwanted sounds—and the role of context, often geographically defined, in shaping human perception and experience of the acoustical environment. But at the same time, the soundscape approach is still highly acoustical.

The meaning people give to sounds and noise, or to an exposure context, strongly influences their reactions and accompanying health effects. Sounds produced by humans can be considered as a product of behaviour, which in its turn is a consequence

Table 1.1 Noise control approach versus the soundscape approach

Noise control approach	Soundscape approach
Concerns sounds of discomfort	Concerns sounds of preference
Measures integrated sounds	Differentiates between sound sources (wanted unwanted)
Manages by reducing levels	Manages by wanted sounds masking unwanted sound
Sound as a waste	Sound as a resource

Source Brown (2010), permission from author

of human needs. As a result of these needs, people produce sounds and expose themselves to sounds, partly with a purpose, partly as an unintended byproduct of their activities. The meanings of these sounds can be negative and positive, they are partly generic, partly culturally defined, and they change over time. The noise control approach has been focused exclusively on the negative aspects and meanings of noise. But in order to understand the driving forces producing noise it is important to study the positive aspects as well. The shift from decibels to perception and meaning in their context offers important cues for future research and policy.

That the appraisal of sound is subjective does not mean that the effects of sound are subjective as well. Both the pleasure and the annoyance one gets from sounds are objective, measurable and real. A misconception often encountered is that differences in the appraisal of sound mean that the experiences of some people are not real. If two people live adjacent to each other and the person living in house number 15 is severely annoyed by road noise while the number 17 residents do not care a bit, some draw the conclusion that the annoyance of number 15 is subjective. The experience of both is real and objective. The ecstasy of somebody dancing in his garden while listening to Ziggy Stardust played at maximum volume as instructed by David Bowie himself, is as real as is the abhorrence of the neighbor who tries to relax in the adjacent garden and prefers to listen indoors to Michael Bublé at mid volume.

1.7   Noise as Silent Killer

Chronic, unwanted and or harmful sounds can make you ill, directly or via a process of annoyance, lack of sleep and sustained stress responses to something you have no control over. There is however strong evidence that direct physiological responses happen independent of annoyance, so also with wanted sound and especially during sleep in adults and children. Although we tend to focus on research as well as policy on high noise levels there is increasing evidence that also at low levels damage can be done. Moreover, the number of people exposed to these lower levels is much larger. From a public health point of view therefore they are more interesting when it concerns preventive measures.

The health effects of high and chronic levels of unwanted sound from transport have been well documented in the past 35 years. The latest WHO noise guidelines

(World Health Organization 2018) for the European region, its underlying reviews and subsequent updates underpin earlier conclusions that environmental noise affects not only long-term well-being and sleep quality, but also cardiovascular disease, cognitive effects, and metabolic effects such as obesity and diabetes type 2. Less clear is the often assumed association with psychopathology, which is not confirmed nor is a causal association plausible, except for (nonclinical) effects as well-being. It could very well be the other way around, that people with mental health issues, be it temporary (state) or innate (trait), are more affected by noise via noise sensitivity. It is well known that people who are depressed or suffer from schizophrenia, and people who have undergone surgery and use certain medications are extremely sensitive to noise. Much less frequently studied are the positive effects or restoring effects of wanted sound or variation of pleasant soundscapes.

1.8   Sound, Noise, Happiness and Health

The focus of the previous paragraphs was primarily on noise, the negative side of sound. It was already mentioned that in daily life the positive effects of sound are predominant. We actively seek pleasant sounds by listening to music or the singing of birds. Familiar sounds of people talking quietly or children playing we continuously hear without consciously noticing make us feel safe and at ease. Sound is a major pathway to experience the world in all its facets.

Our first sensory experience of the world is through sound. A foetus starts to hear sound around 20 weeks; hearing the reassuring heartbeat of the mother and establishing the first contact with the outside world. These first sounds are pleasant sounds. They are not pleasant because they are entertaining, but because they signal safety and familiarity, literally and proverbial. This capacity has even been targeted commercially, by selling products that are supposed to boost the baby's intelligence even before it is born (https://www.amazon.com/Nuvo-Ritmo-Pregnancy-Sound-Sys tem/dp/B002XFE89). From a very early age, music enters the scene, not only to soothe (or increase intelligence) as can be done prenatally. Babies from 5 months old can react to rhythm with dancing movements and seem to enjoy it very much (https://www.sciencedaily.com/releases/2010/03/100315161925.htm). This hints at the prehistoric roots of music with the human voice and instrument playing a role in the upbringing of children, to impress other tribe members from the same or other sex, to lure or deceive prey, in religious practices and other uses.

A pleasant soundscape at home and in the residential environment is according to some good for health, but evidence is still limited. With growing traffic and populations in urban areas, fewer people get access to such soundscapes.

1.9   Urban Planning

The bridge between soundscape and urban planning is easily made. Well-being and health in the urban setting are topics that logically are or should be the focus of urban planners, architects and other professionals who play a role in building cities. Being a complex topic, it is not surprising that there are widely differing views on what constitutes a 'good and healthy' city. We saw already that sound constitutes an important component of city life, but acoustic design has been vastly neglected in modern Western architecture. Views on urban sound vary within urban planners and architects. Generally, laypeople and professionals do not have much awareness about the intersection of sound and architecture. Nevertheless, acoustic design is an essential part of a successful design of a building. As stated by Max Neuhaus (https://www.kun stradio.at/ZEITGLEICH/CATALOG/ENGLISH/neuhaus1-e.html), "The ear does things which the eye can't do; the eye does things which the ear can't do. In addition, visual and aural perception are complementary systems. It is not a question of one being better than the other; they fit together".

Two influential people in contemporary thinking about Western urban planning were Jane Jacobs and Le Corbusier, representing on the one hand the diversity of the open city concept versus homogeneity as the golden standard in an industrial view of building (see Figs. 1.2 and 1.3). Sound played a back role in their opinions.

Following the large-scale introduction of trams, railways and other machinery noise became a societal concern in the Western world in the early 1900s and anti-noise movements arose in several countries (see Chapter on governance). Explicit attention to the impact of traffic in relation to urban planning was provided by Appleyard (1980) from the 1970s onward and later by his son Bruce, whose ideas show some

Fig. 1.2   Sketch of New York Streets by Jane Jacobs (Dover and Massengale 2013) with permission

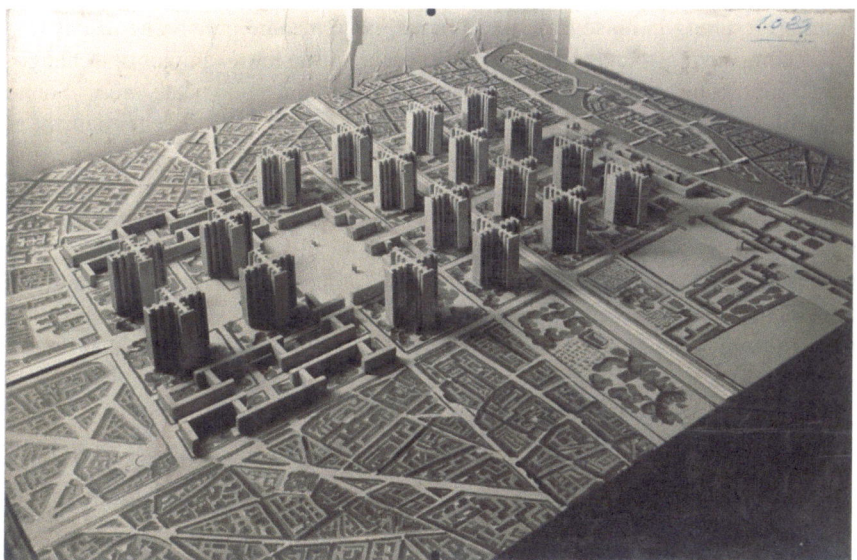

Fig. 1.3 Ideal city according to Le Corbusier (1933), © FLC/ Le Corbusier, La Ville Radieuse c/o Pictoright Amsterdam 2024

similarities with those of Jacobs in the sense of a more social and political view on planning in which equity is a dominant aspect. This was a response to the dominant view on transport, also nowadays, as efficiency is at the core of the transport system. They made a plea instead for livable streets and showed convincingly how people's interactions and communications, and their sense of belonging are strongly affected by the infrastructure, e.g. the presence of a main road dissecting the neighbourhood creating a barrier and potentially socio-economic inequalities. Instead of putting modes of transport, mostly cars, in the center of attention, people should be the central point of attention. This is relevant for developed countries and essential for developing countries where inequalities are larger and the impact of transport is larger.

1.10 Is Noise Only a City Problem?

So far we only spoke of urban planning, as if the rural area does not exist. But sound is important everywhere and not exclusively in cities. Cities are crowded by definition and from the beginning of urban history cities are the centers attracting people for many reasons, often economic. Cities are densely populated, need continuous supply and digest enormous numbers of daily visitors. All of this implies noise. Rural areas have their own range of noises and the rural soundscape is rapidly changing. The noise from large wind parks and more recently the scaling up and industrialization of farming, resulting in noise from mega farms, large manure processing installations and accompanying sounds of freight transport are just a few examples of a changing rural soundscape. Major roads often run through thinly populated areas. Sounds from all these traffic and industrial sources scattered over the rural environment are more salient and noticed since the background noise levels are much lower than in the city. As indicated above, climate change related measures such as increased freight transport via rail during the night is in particular problematic in rural areas for the same reasons. The same counts for the evidence of a relative increase of low-frequency noise, which again is more noticed in quiet areas and propagates over larger distances. A growing number of people are disturbed by an often unidentified low-frequency sound that some scientists now call the worldwide hum (https://www. thehum.info/).

1.11 Noise Governance

It was at the end of the 19th and the beginning of the twentieth century that noise and noise annoyance became a societal concern in the Western world. Anti-noise movements arose in different countries such as the US, Australian, and European cities (https://noisenewsinternational.net/the-first-international-anti-noise-con ventions-congresses-1895-1912/). These movements were focused mainly on local problems. For some, this was for religious reasons, wanting a quiet environment to pray to God, while for others the major issue was sleep disturbance and general health from increased stress.

One of the world's first documented noise abatement societies was formed in London in 1895: the Association for the Suppression of Street Noises (ASSN). Many cities followed. It was preceded and accompanied by a strong wish to measure sound, something musicians and scientists had dreamed of and tried for centuries. In 1878 Edison developed a phonograph able to measure the noise produced by the Metropolitan Elevated Railway of New York. In 1895, the inventor Hiram Maxim proposed "to measure" the sound level in a legal dispute about noise emission from an electric power plant. In 1905, in Boston, a phonograph recording was admitted as evidence in a US court in a case against the Boston Elevated Railroad. This case had a big impact. These examples show how closely related noise abatement, noise policies and sound levels are, up to the present day.

One of the first environmental noise surveys was performed in New York in 1926 (Rosen 1974) and many followed. They provided the basic ingredients for health impact assessment. Health Impact Assessment (HIA) is in current times increasingly used in the development of environmental and public health policies and regulations. It commonly involves the identification of environmental hazards, and the quantification of the expected Burden of Disease (BoD) and accompanying costs. This (environmental) disease burden can be expressed in a variety of ways. WHO and others increasingly use the DALY: Disability Adjusted Life Years, encapsulating the amount of healthy life years. The long-term effects of transport noise are now reasonably well described for several health outcomes and dose-response relations have been established for them (World Health Organization 2018).

After selecting a set of endpoints with sufficient evidence for a relationship with the risk factors under study, the expected environmental disease burden in a population can be quantified by combining population-density data with concentration distributions of a relevant exposure indicator, information on exposure-response relationships, and estimates of the duration and severity of diseases. The DALY is in the first place a relative measure comparing the impact of noise with other exposures such as air pollution, radon, UV radiation, and damp houses and compared at the national level. The calculation made in 2010 for the year 2020 by Knol en Staatsen (2005) for the Netherlands for instance predicted that the disease burden associated with noise exposure would increase up to a level similar to the disease burden attributable to traffic accidents.

1.12 Noise and Power

As repeatedly stated, sound experiences are subjective and only people themselves can decide if a sound is wanted or unwanted and if it counts as noise or not. When sounds and noise become a societal concern government decides which sound levels count as noise and when the risk of adverse effects becomes unacceptable. This is somewhat comparable with the way we deal with most health risks. Measures are applied to all, while only a fraction runs a risk. This is often referred to as the prevention paradox (Thompson 2018) stating that the majority of cases come from a population with low or medium risk simply because those groups are larger; we are all preventing diseases that we would never get in the first place. But at population level that works.

Decisions on hazards are made in an arena where people fight for attention from decision-makers who have their own backgrounds and biases.

The opposition to industrial noise in the second half of the nineteenth century culminating in the establishment of the noise abatement societies was led by affluent people who preferred tranquility. So decisions on noise are inextricably linked with issues of power and control. A striking example is the war on noise of New York City Mayor Fiorello LaGuardia in the 1930s (https://thelocal.to/how-noise-shaped-a-city/). According to Lilian Radovac, LaGuardia "approached noise as an aural barometer of the chaos of New York City", both a symptom and cause of urban disorder. He "brought sound to the forefront of a wide-ranging campaign to rehabilitate, reorganize, and transform urban life". Noise complaints in New York were used as a pretext to hassle the black and immigrant residents of gentrifying neighbourhoods, to disrupt political protests and strikes, and to get merchants off the sidewalks and children out of the streets.

Similar endeavours were performed in many US and European cities in the name of noise control since the second half of the nineteenth century.

This continues to the present day as is seen for instance in the discussion about the tranquility that the lockdowns to reduce the spread during the coronavirus pandemic in 2021/2022 brought, highly appreciated by some and detested by others (https://www.theatlantic.com/technology/archive/2020/07/the-str uggle-for-the-urban-soundscape/614044/). The latter literally broke the silence with loud protests and nightly fireworks which prompted reactions from the authorities struggling with the situation.

Noise complaints are filed more often by homeowners than renters, more by the higher socioeconomic classes than the lower, more by suburbanites than urbanites. At the same time, in many cities the poor are exposed to higher traffic noise levels, louder neighbours and other noisy sources than the better off.

1.13   Outline of the book

In this book, we hope to have covered all topics important to understanding the wide-ranging implications of the health effects of sound and briefly touched upon in this introduction. Its main aim is to make clear that adequately considering the effects of sound is more than calculating dose-response relationships between decibels and specific health effects. Sound has to be studied from its social, cultural and physical context. Our ideas and judgements about ambient sounds are as much the result of decibels as of the attitudes we have towards their sources. Meanings and perceptions are as important as the physical characteristics.

Historically most attention is devoted to the negative health effects of unwanted and source-specific sound. The most recent noise guidelines (World Health Organization 2018) for the European region of the World Health Organization illustrate this well.

In this book, we try to counterbalance this focus by placing the issue of noise and health in a broader societal context. It describes the current trend of considering sound not merely as an unwanted waste product, but also as an asset to be used in urban planning. It describes next to the negative health impact of noise exposure, the beneficial effects of high acoustic quality. Methods to quantify and summarise the impacts are described and related to a broader environmental context and their economic effects. Critical to the discussion is to consider how sound has been dealt with in policy and legislation and what interventions work and have potential but are ignored in noise policies, what are the drivers and obstacles for noise abatement measures? And what noise- and sound-related problems are to be expected in the future and how can we deal with them?

The approach is highly interdisciplinary combining knowledge and insights from the physical, medical, psychological, economic, spatial planning, governance and art disciplines.

This book is not about acoustics, but inevitably uses terms and concepts common to the world of sound and noise. In Chap. 2, Calum Sharp provides an explanation of the acoustical terms, concepts and methods which are used to measure, quantify and understand sound. It includes a glossary with the definitions of the most important terms. In just a few pages, Calum Sharp reveals the main landscape of the study of noise and sound and its vocabulary. Especially to the reader outside the world of acoustics, this introduction is a helpful guide for the whole of the book.

The undesirable health effects of noise as they are unraveled in, for instance, the WHO environmental guidelines are most often summarized in dose-response relationships. The dose in this is always the decibel and the dose-response relationship is between the sound level of a given source and a certain effect. For this sound level a measure is taken, like L_{den}. This method presumes that the sound level is the most important determinant of the given health effect. In reality, health effects are determined by a multitude of acoustical and non-acoustical interplaying factors. The non-acoustical factors can be personal (fear, noise sensitivity, attitudes), social (socioeconomic status SES, cohesion), situational (crowding) and cultural (individualist versus group culture). The effects of sound are largely dependent on the meaning a noise has for the receiver and not on the sound itself. Music, for instance, can be used to lift your spirits and to reach a state of ecstasy or it can be used in torture as was done with the prisoners in Guantanamo Bay. Rainer Guski explains all this in Chap. 3 in detail and also hints at what this means for noise abatement and sound policies in general.

In Chap. 4 Charlotte Clark, Danielle Vienneau and Gunn Marit Aasvang give us the state of the art on noise and health. It is not meant as a systematic review, but as an overview showing the health effects with its key sources, summarizing the basic facts with reference to the latest evidence. Next to this basic information, the Chapter goes into the strong and weak points of the approaches that have been used to determine health effects, the uncertainties in it and the role of specific groups and places. The focus in this Chapter is traditionally on the negative health effects of 'noise', but some space is devoted to the positive health effects of sound.

Chapter 5 puts the adverse health effects of sound in a broader context. Juanita Haagsma and Mark Brink explain how the health impact of environmental noise can be summarized in measures of health burden like a DALY or QALY. The results of this can be used to compare the health burden of noise to other environmental exposures and other determinants of health. DALYs can be monetarized and this can be used as input to a cost-benefit analysis. DALYs are widely used and can be very helpful in making policy decisions. At the same time, they have their drawbacks and risks.

In Chap. 6 Ronny Klæboe looks at several economic aspects of noise and health. Economic development has been a driver of noise since ancient history. It was the reason why the Romans let their carts drive at night. In the present day, economic aspects are part of noise policies directly. As was mentioned above, monetarized DALYs can be used in a cost-benefit analysis. Some countries also use calculations to decide upon the cost-effectiveness of abatement measures. In the Netherlands, for instance, an efficiency criterion is used to determine if a noise barrier is built along highways. It is only built if a certain number of dwellings will have a reduction of a certain amount of decibels according to a prediction model. From what is discussed in previous Chapters it is clear that such a calculation does not take all considerations into account. It uses the premise that the effects completely depend on the noise level. Economic aspects determine the willingness and possibilities to take measures to a large extent. An important reason why noise annoyance is so difficult to combat is because it is very hard to solve noise problems technically, as can be more easily done with air pollution. Noise measures are often difficult to implement, take up space and are costly.

Zoning may be one of the most difficult to implement and costly noise abatement measures. In zoning, noise considerations lead to identifying spaces as not suitable for often profitable functions like housing. In the city, this can mean that space for building houses can be unavailable because of noise contours. This also shows the interplay between sound and spatial planning. Next to zoning, it includes the orientation of buildings, functions of buildings, (socio-cultural) composition of neighborhoods, stretching to the idea of acoustic (instead of visual) planning and designing into the soundscape approach sometimes accompanied by sound art. It links to the visions of the urban planners mentioned above with sound as an important building stone of the city. Trond Maag and Arnthrudur Gisladottir discuss these approaches in Chap. 7 with their merits and limitations. It goes beyond the idea of fountains masking traffic noise. It looks at the role of sound (both negative and positive) in the built environment and what can be done with planning to make especially urban environments acoustically as pleasing as possible.

It is often said that air pollution is the most important environmental burden because everybody has to breathe air. In food, drinks, soil and others the exposure to chemicals can be prevented. If the inevitability of exposure is the criterion, noise surely ranks at least as high as air pollution. More than air pollution, sound also has its positive side. We are immersed in a sonic environment from which we cannot escape. Marcel Cobussen in Chap. 9 claims that people are entitled to a healthy sonic environment. To achieve this, he takes urban sound planning a step further in his Chapter on sonic ecology and the role of sound art in creating a healthy urban environment. Artists, working together with residents, urban planners, project developers, and (local) officials, can use their creativity to design places where people have more attentive interactions between human beings and their sonic environment. By doing this, the sonic environment can contribute to belonging and connecting and add quality to the public space with a positive effect on its users. For Cobussen, artists offer a major contribution to creating a positive sonic environment. This does not mean they make everything more beautiful as some would expect from artists. Their objective is to enhance the possibilities for human as well as non-human agents to affect and be affected. Cobussen illustrates his ideas with a project in his hometown Rotterdam he was involved in.

Many countries have noise policies and governance practices on noise. Benjamin Fenech and Natalie Riedel discuss governance issues on a global, continental and national scale in Chap. 8. Because practices differ widely between countries, the examples are limited and somewhat selective. Key in this Chapter are the different approaches that have been used in different parts of the world to minimize the health effects of noise (if taken at all) and how successful these have been. This success is closely related to changes in noise exposure which has been increasing in many parts of the world. Noise measures can be like a Sisyphus task, having to roll a heavy rock up a steep slope where it inevitably slides down sometimes.

The governance practices discussed in Chap. 8 often contain regulations and measures at the sources of noise (quieter airplanes and road surfaces, for instance), on the propagation of noise (noise barriers along roads for instance, but also zoning) and on the exposed public (insulation of houses). Measures can also be taken in public space, like providing access to quiet places or focus on communication or the establishment of good relationships between the producers and the receivers of noise. Measures are often directed towards noise levels and less frequent on the many non-acoustical factors discussed by Rainer Guski in Chap. 3. Measures can also be effective if they influence the non-acoustical factors. Be they acoustical or non-acoustical, a basic requirement of a measure is that it is effective.

The world of sound and health has a rich history, but also has a challenging future.

Future projections of noise exposure have been taken into account in many governance practices and policy documents from the early days of noise governance. Often, these projections were not borne out. In the Netherlands, traffic noise was allowed to be 5 dB higher than the limit value for a period of 10 years because it was expected that traffic would become 5 dB quieter in that period. When this did not happen, the limit value was raised 5 dB, with the legitimization that is had been common practice for a long time. Noise-producing machinery (cars, air planes, factories) have indeed become quieter, but this is offset by a growth in volume of machines and in unexpected increases in noise exposures like the noise from electric car tyres. It cannot be counted upon that noise levels, especially in cities, will drop, but the sound landscape will certainly change in the future.

In Chap. 10, Antonio Torija Martinez looks at future developments in the technology of transport modes, the effects of the energy transition (wind turbines, noise of ventilation systems, heat pumps, cooling systems) and in changes in society having an effect on noise exposure like individualization, spatial planning, changing expectation of people, the role of quantified self and citizen science.

In the concluding Chap. 11, the editors Irene van Kamp and Fred Woudenberg discuss the main findings of the book. They draw up the balance sheet about what is known about sound and health and what the book teaches us on how we can move on to reduce the negative effects of sound and increase its positive effects.

References

Appleyard D (1980) Livable streets: protected neighborhoods? Ann Am Acad Pol Soc Sci 451(1):106–117

Bloem JC (1946) De Dapperstraat. In Quiet though sad: poems. A.A.M. Stols, 's-Gravenhage, The Netherlands

Brown AL (2010) Soundscapes and environmental noise management. Noise Control Eng J 58:493–500

Coates PA (2005) The strange stillness of the past: toward an environmental history of sound and noise. Environ Hist 10(4):636–665. The University of Chicago Press

Dover V, Massengale J (2013) Street design: the secret to great cities and towns. Wiley

Duarte MHL, Vecci MA, Hirsch A, Young RJ (2011) Noisy human neighbours affect where urban monkeys live. Biol Lett 7(6):840–842. https://doi.org/10.1098/rsbl.2011.0529

Grinde B, Patil GG (2009) Biophilia: Does visual contact with nature impact on health and well-being? Int J Environ Res Public Health 6(9):2332–2343. https://doi.org/10.3390/ijerph6092332

Hempton G, Grossmann J (2009) One square inch of silence: one man's search for natural silence in a noisy world. Simon and Schuster

http://www.antiquitatem.com/en/ancient-cities-and-noise-deafness-sybari/

https://noisenewsinternational.net/the-first-international-anti-noise-conventions-congresses-1895-1912/

https://www.amazon.com/Nuvo-Ritmo-Pregnancy-Sound-System/dp/B002XFE89

https://mikegoldsmith.weebly.com/history-of-noise.html

https://ourworldindata.org/urbanization#:~:text=More%20than%204%20billion%20people%20live%20in%20urban,world%20lived%20in%20urban%20than%20in%20rural%20areas

https://thelocal.to/how-noise-shaped-a-city/

https://www.aeraweb.org/blog/2009-field-season/the-sounds-of-antiquity/

https://www.asor.org/anetoday/2015/08/policemen-in-1st-millennium-bc-babylonia/

https://www.byevisa.com/tomorrows-top-remote-working-destinations/

https://www.kunstradio.at/ZEITGLEICH/CATALOG/ENGLISH/neuhaus1-e.html

https://www.nhm.ac.uk/discover/news/2022/july/underwater-noise-pollution-risking-lives-whales-dolphins.html

https://www.sciencedaily.com/releases/2010/03/100315161925.htm

https://www.theatlantic.com/technology/archive/2020/07/the-struggle-for-the-urban-soundscape/614044/

https://www.thehum.info/

Hu Y, Goncalo CC (2009) Are bird species that vocalize at higher frequencies preadapted to inhabit noisy urban areas? Behav Ecol 20(6):1268–1273. https://doi.org/10.1093/beheco/arp131

Knol AB, Staatsen BAM (2005) Trends in the environmental burden of disease in the Netherlands, 1980–2020 RIVM report 500029001, Bilthoven, The Netherlands

Le Corbusier (1933) La ville radieuse, Editions de L'Architecture D'Aujourd'Hui, Boulogne (Seine)

Netherlands Health Council (2020) Health and the energy transition in the built environment. Netherlands Health Council, The Hague

Rosen G (1974) Public health, then and now. A backward glance at noise pollution. Am J Pub Health 64(5):514–517

Simona J, Rossell JM, Sánchez-Roemmele X, Vallbé M (2021) Evolution in the law of transport noise in England. Transp Environ 100(special issue Transport Research part D: Transport and Environment). https://doi.org/10.1016/j.trd.2021.103050

Thompson C (2018) Rose's prevention paradox. J Appl Philos 35(2):242–256. https://doi.org/10.1111/japp.12177

Ulrich RS (2008) Biophilic theory and research for healthcare design. In: Biophilic design: the theory, science and practice of bringing buildings to life, no 1, pp 87–106

van Kamp I (2022) Energy transition related noise and vibration issues and its health consequences. In: Proceedings internoise 2022, August 21–24, Glasgow

Waller S (2019) What our prehistoric ancestors can teach us about echoing soundscapes that is relevant to modern noise studies. J Acoust Soc Am 146(4):2826–2826. https://doi.org/10.1121/1.5136790

Wilson EO (1986) Biophilia. Harvard University Press

World Health Organization (2018) Environmental noise guidelines for the European Region. World Health Organization, Copenhagen

Wunderli J-M, Brink M (2022) Long-term evolution of noise annoyance depends on the type of transportation noise—What are the main drivers for the observed trends and their differences? In: Proceedings internoise 2022, August 21–24, Glasgow

Chapter 2
Introduction to Acoustics: Measuring, Quantifying, and Understanding Sound

Calum Sharp

Abstract To understand sound, noise, and their effects on health, we must quantify sound using the concept of 'acoustics'. This chapter provides an introduction to acoustics and outlines different ways that acousticians measure, predict and quantify sound to understand its effects. The chapter also contains a glossary of acoustic terminology that is used throughout this book and can be referred to when reading any of the other chapters in the book.

2.1 Introduction

> If a tree were to fall on an island where there were no human beings would there be any sound? Chautauquan (1883).

This quote from an 1883 magazine article is thought to be the first time that the popular question (often posed in the alternative form "if a tree were to fall in a forest...") was written. While the question is now more often raised as a thought exercise, the article did provide a definitive answer:

> No. Sound is the sensation excited in the ear when the air or other medium is set in motion.

This answer is provided by the consideration of the science of sound—acoustics. As Rainer Guski notes in Chap. 3, the sound of a tree falling on an island or in a forest, along with all other sounds in the world, are just vibrations in the air first caused by physical movement. These vibrations, or 'soundwaves', travel through solid mediums and through the air until they reach and vibrate our ears, sending electrical signals that our brains then decode as sound.

Though it is tempting to feel that we have lost some of the mystery in the question by answering it in such a definitive manner, it is wonderful to consider that all the sounds in the world, everything from birdsong to the strings of an orchestra, are created and propagated by simple vibrations, each one with a unique pattern.

C. Sharp (✉)
Arup, London, UK
e-mail: Calum.Sharp@arup.com

Fig. 2.1 Image of a waveform

Capturing and recording these vibrations is key to the concept of acoustics. To understand sound, we must measure and quantify it. Acousticians will use microphones and specialist recording equipment known as Sound Level Meters, as well as their own highly trained ears, to do this.

Our ears, microphones and other forms of recording equipment all work in a similar manner - by providing a moveable medium that can be vibrated when the vibrations of a sound reach it. In the case of our ears, this is our eardrum. In the case of a microphone it is a specialist membrane within the device alongside a mechanism that converts the vibrations into a signal that can be stored and recorded.

The recording of a sound in its simplest sense, is a representation of the vibration that occurs in the air and in our eardrums over time when we hear the sound. The peaks of the waveform, or the 'amplitude', represent the relative loudness of the sound, and the width of the waveform between peaks and troughs represents the 'frequency' or the 'pitch' of the sound. In most cases, unless the waveform is magnified many times, it is not possible to visually interpret the frequency of a waveform, because most sounds contain frequencies that complete a cycle several thousand times in one second. Frequency is measured using the unit Hertz (Hz), named after Heinrich Rudolf Hertz. 1,000 Hz represents 1,000 cycles per second (Fig. 2.1).

Playing back a recorded sound is essentially the same process but reversed. A music playback system will start by reading and interpreting the waveform of a piece of recorded sound (for example from the physical grooves in a vinyl record or the digital data of an mp3) and then sending that signal to a speaker (which may be a large free-standing loudspeaker, or a tiny speaker inside a mobile phone or a pair of headphones). Most loudspeakers contain one or more cone-shaped diaphragms which vibrate backwards and forwards in the same 'shape' as the recorded waveform, ultimately attempting to replicate the vibration that caused the original sound.

Again, it is wonderful to think that every recorded sound in the world can be reproduced to some extent by the simple forward and backward vibrating motion of a loudspeaker.

Fig. 2.2 A Sound-Level Meter and examples of decibel levels from different sound sources

2.2 Measuring Sound

2.2.1 Sound-Level Meters

As well as the visual representation of a waveform, there are many technical aspects of a sound that can be recorded and measured. One of the most common items of equipment used by acousticians is the Sound Level Meter or SLM. An SLM is a highly precise and calibrated microphone attached to a small computing system that can measure particular aspects of a sound. Whilst SLMs and their associated equipment have varied in size over the history of their use, SLMs are now generally small enough to hold in the hand or mount on a camera tripod (Fig. 2.2).

2.2.2 Sound Pressure Level

SLMs are capable of recording a wide variety of sound information or 'metrics'. The most common and simplest metric is the Sound Pressure Level (SPL). This is a measure of the pressure fluctuation caused by sound vibrations relative to a reference value and is related to the perception of the loudness of a sound—a louder sound will have a higher SPL.

The unit used to measure SPL is the decibel (dB), named after Alexander Graham Bell. As the range of sound pressures that the human can hear is huge (the sound of a jet engine creates sound pressure levels around 1,000,000,000,000 times greater than a sound that is just audible), the dB uses a logarithmic measurement scale. This means that it cannot be interpreted linearly, i.e. a doubling of the sound pressure level does not result in a doubling of the SPL, but rather a 3 dB increase. It might be intuitive to think that a doubling of the SPL may also result in a doubling of the perceived loudness. However, it is generally accepted that a 3 dB increase is around the point at which humans are able to perceive the difference in loudness between two sounds. It is not until you reach a 10 dB increase (representing a tenfold increase in SPL), that sounds are typically perceived to be twice as loud. This is of course a rule of thumb and an oversimplification, as most of the chapters in this book note the perception of sound will vary greatly depending on the source of the sound, its acoustic character, the context in which it is experienced and the attitude and sensitivity of the listener.

2.2.3   A-weighting

SPL, and other metrics which utilise the dB, are usually accompanied by a 'weighting' which affects the extent that different frequencies (recalling that frequency represents the pitch of a sound) will contribute to the measurement. The most typically used weighting is the A-weighting, which attempts to replicate the sensitivity of the human ear to different frequencies. Humans are relatively less sensitive and able to perceive low and high frequencies (typically below 20 Hz and above 20,000 Hz). Applying A-weighting therefore suppresses the extent that low and high frequencies contribute to the measurement, giving more prominence to the audible frequency range. When A-weighting is applied to a SPL measurement the resulting metric is typically represented as dB(A). Other metrics will include the A as a subscript.

2.2.4   Quantifying Sound Exposure

While SPL is a useful measure of the loudness of a sound, it is an instantaneous metric which captures a snapshot of noise data from a particular point in time. It therefore has limited use in quantifying a sound for the purposes of determining human response and health implications which are generally influenced by longer term exposure.

There are a number of metrics which take into account the varying SPL of a sound over a set time period, which helps to quantify the longer term 'dose' of exposure to sound. One such metric is the $L_{Aeq,T}$ metric or 'equivalent continuous sound level'. This is a measurement of the total sound energy over a period of time, T. It is easiest to think of this as an average, but important to note that all the sound energy in the time period is captured by this metric. In the United Kingdom, L_{Aeq} is typically measured during the daytime period (07:00—23:00 denoted $L_{Aeq,16h}$) and the night-time period (23:00—07:00 denoted $L_{Aeq,8h}$). For most metrics relating to sound and health, it is common to measure and quantify daytime and night-time exposure separately, noting that the sensitivity to sound is different during these periods, and the health outcomes are different too: for example, annoyance during the daytime and sleep disturbance during the night-time.

Another common metric for measuring long-term noise exposure is the L_{den} metric, where 'den' refers to 'day, evening and night'. This metric is predominantly used in mainland Europe in line with the Environmental Noise Directive of the European Union. The L_{den} is a 24-h metric which captures noise across the day (07:00–19:00), evening (19:00–23:00) and night (23:00–07:00) periods.[1] The L_{den} metric provides greater emphasis on noise exposure during the evening and night, applying a 5 dB and 10 dB penalty respectively, recognising the greater sensitivity to noise during these periods.

While long-term exposure metrics are important for understanding health exposure, instantaneous metrics which quantify the peak level of a sound event are also valuable, particularly during the night-time, when individual loud events can result in sleep disturbance and awakening from sleep. The simplest measure of the highest energy of a sound event is the L_{Amax} metric, which aims to capture the loudest part of a sound event. The LAmax metric will often be accompanied by an 'S' or 'F' denoting whether a slow (0.125 s) or fast (1 s) time window is used to calculate the peak of the sound event. By itself, the L_{Amax} can provide limited information, as it only provides information on a single sound event, but when the L_{Amax} of all sound events over a time period (typically the night-time) is captured it can provide useful information that can be used to relate to the probability of being sleep disturbed and/or awoken during the night.

Although all of the above metrics are useful in quantifying long-term health exposure, as is noted in most of the chapters in this book, there are limitations to metrics that quantify simple noise exposure using metrics that are ultimately based on A-weighted sound pressure level. Such metrics capture the sound energy, but do not provide any information on the character or context of a sound. These aspects are particularly important for new or novel sounds resulting from technological advancements such as electric vehicles, drones and novel aircraft as discussed by Antonio Martinez in Chap. 10 of this book. As noted in that chapter, research is ongoing to develop metrics to account for unique sound signatures generated by new sources of sound.

---

[1] Note the definition of the day, evening and night periods can vary from country to country.

2.2.5 Understanding Sound Exposure

As noted in Chaps. 1 and 3, noise is generally defined as 'unwanted and/or harmful sound'. Convention, therefore, is therefore to use 'sound' when referring to objective physical aspects of a sound and 'noise' when referring to adverse subjective psychological or physiological responses to sound.

Examples of 'sound' terminology	Examples of 'noise' terminology
Sound level	Noise impact
Sound exposure	Noise mitigation
Sound emission	Noise control
Sound predictions	Noise effects

Transportation sound is generally considered noise (those that enjoy or welcome the sound of transportation in their day-to-day lives are few and far between). It is therefore common to use 'noise' in terminology when referring to transportation sound.

In order to understand the effects of transportation noise (or any other source of noise), it is important to quantify the exposure to noise over time, for example using the metrics described earlier. This can be achieved by measuring the sound— using one or multiple sound level meters. However, this approach can be limited as it will only provide information on the sound exposure at the specific measurement location. It is also generally not possible or straightforward to separate out the exposure associated with different sound sources, which in many cases, particularly in urban environments, could be a combination of several different continuous and instantaneous sources of sound.

When it is important to quantify sound exposure over a larger area, or to be able to distinguish the effects of different sources of sound, the typical approach is to model the sound. There are several computer software packages designed to be able to do this—taking into account the way a source will emit sound (a combination of the 'power' of the sound, as well as the direction or 'directivity' in which the sound will travel), and the way sound travels, reflects and is absorbed as it moves through the air or when it comes into contact with surfaces (which may or may not be sound absorbing). A good computer sound model will also be checked or 'calibrated' using actual measurements, to ensure that the outcome of the model is a reasonable reflection of reality. This model validation is an important part of the process of quantifying and understanding sound exposure.

Once the sound exposure is quantified, it is then possible to determine the potential adverse effects as a result of the exposure. When new sources of sound are introduced that may affect noise-sensitive receptors, there is usually a requirement in planning policies to undertake an assessment of the noise exposure and its effects and introduce mitigation to minimise those effects. There is a myriad of different methodologies that can be employed to do this. Most sources of transportation have their own methodologies for modelling and predicting sound, and for quantifying adverse effects from noise. Different countries may also have their own methodologies. In many cases, the methods involve thresholds of noise above which adverse effects can be detected and separate higher thresholds above which *significant* adverse effects can be detected. These thresholds are typically informed by research studies such as those described in Chap. 4. As well as the absolute noise exposure relative to certain thresholds, it is important to quantify the 'change' in sound exposure as a result of a new sound source—a larger change in sound is more likely to result in adverse effects, potentially even at lower absolute levels of exposure.

2.2.6 Controlling and Mitigating the Effects of Noise

Once the potential adverse effects of noise are understood and quantified, the next stage may be to control and mitigate, or even avoid, these effects. Noise mitigation typically follows a mitigation hierarchy:

1. Mitigation at the noise source itself,
2. Mitigation of the source-to-receiver path,
3. Mitigation at receptor (where the receptor is the person, building or location that is potentially exposed to the noise).

Mitigation at the source comes first in the hierarchy and is usually the most effective, as it reduces the noise of the source, which benefits all receptors. When a good acoustic design is employed early in a design and optioneering process, then significant benefits can be had by routing transportation corridors or other sources of noise away from noise-sensitive receptors. Once the position of a noise source is fixed, mitigation at the source could involve the use of quieter engines for aircraft or low-noise surfacing for roads. For construction noise sources it could involve the selection of quieter equipment or putting generators and other noise-generating equipment inside enclosures.

The second step in the hierarchy is to provide mitigation between the source and receiver. Again if good acoustic design is employed in urban planning, as discussed in Chap. 7, then masterplan layouts and buildings can be used to shield the more sensitive parts of a development from noise. Along transportation corridors, it is common practice to employ noise barriers. A noise barrier can be any solid structure that breaks the 'line of sight' between the noise source and the receiver, such that the sound energy has to pass through the barrier (where some of its energy will be reflected and some absorbed) or travel around the edges of the barrier (a phenomenon known as 'refraction'). Barriers can be simple vertical structures made of concrete or timber, or complex landforms such as bunds and earthwork cuttings. Noise barriers are most effective when they are positioned close to the source or close to the receiver. They are therefore particularly effective for railways where the barriers can be positioned very close to the trains. Noise barriers are also very common along the side of highways, though they can be less effective for roads with multiple lanes as the barrier ends up being further away from part of the noise source (vehicles on the furthest lanes).

Finally, the last step in the hierarchy is to mitigate at the receiver. Such a step should only be employed where mitigation at source, and in the source-to-receiver path, has been implemented, and it is no longer cost-effective (see Chap. 6) to provide further mitigation in this form, but a risk of adverse effects remains. In such cases, it is common to provide noise insulation at the receiver. This is undertaken by providing modifications to an existing building that improve its sound insulation (a measure of how effective a building façade is at reducing the sound indoors compared to the sound outdoors).

2.3 Glossary

The following table presents a glossary of acoustic terminology used throughout this book and can be referred back to when reading any of the other chapters in the book.

Term	Meaning
Acoustic ecology	A study of the relationship between human beings and their environment, through the medium of sound.
Archaeoacoustics	The study of the relationship between people and sound throughout history.
Awakenings	An objective measure of sleep disturbance representing when a person is awoken from sleep by noise.
BCR	Benefit–cost ratio. A measure of the economic efficiency of a project or development by comparing the benefits that result with all the costs associated with achieving those benefits.
BoD	The Burden of Disease is a quantification of a health impact measured by financial cost, mortality, morbidity or other factors.
CBA	Cost–benefit analysis. The analysis of all costs and benefits associated with a project or development.
CHERIO	Cumulative Health-based Environmental Risk Indicator. The CHERIO aims to identify locations where (future) residents are at an increased risk due to accumulated environmental exposures. It is the burden of disease expressed in a percentage at a specific location attributable to environmental factors compared to the total burden of disease in the same population and is thus country or region specific.
CTL	Community Tolerance Level. The CTL method assumes that the shape of exposure–response relationships is similar for most communities, but can vary in terms of the "starting point" of the relationship. This variance can be used to determine the difference in 'tolerance' to a sound source for different communities and hence is a way of quantifying the effect of non-acoustic factors that may vary between communities.

(continued)

(continued)

Term	Meaning
DALY	Disability Adjusted Life Year. A measure of the number of years of healthy life that is lost due to health conditions as a result of, for example, noise exposure. One DALY represents the loss of the equivalent of one year of full health.
Decibel (dB)	dB is the measurement unit used for quantifying the sound pressure level. It uses a logarithmic scale.
EIA	Environmental Impact Assessment. An assessment of the significance of the effect of a project or proposal on the environment (including the effects of noise).
END	Environmental Noise Directive. The principal European Union law to identify noise exposure and the means to control it.
Exposure–response functions	Exposure–response functions refer to a relationship between noise exposure and a particular adverse effect (such as annoyance or sleep disturbance) that can be mathematically plotted. These relationships are derived from research studies and can be used to calculate the proportion of adverse health effects that may occur over a population based on their noise exposure.
Frequency	Frequency is the rate of repetition of a sound wave. The subjective equivalent in music is pitch. The unit of frequency is the hertz (Hz), which is identical to cycles per second. A 1000 Hz is often denoted as 1 kHz, e.g. 2 kHz = 2000 Hz. Human hearing ranges approximately from 20 Hz to 20 kHz.
HIA	Health Impact Assessment. An assessment of the health impacts of a project or proposal on the population (including the effects of noise).
ICBEN	The International Commission on Biological Effects of Noise. ICBEN is a scientific organisation whose goal is to promote a high level of scientific research concerning all the aspects of noise-induced effects on human beings and on animals including preventive regulatory measures and to keep alive a vivid communication among the scientists working in that field.
Intermittence Ratio	A measure of the 'intermittency' of sound sources which contain clear distinguishable noise events that are intermittent such as railway and aircraft sound.
$L_{Aeq,T}$	The 'equivalent continuous sound level'. This is a measurement of the total sound energy over a period of time, T. It is easiest to think of this as an average, but important to note that it is a logarithmic average and all the sound energy in the time period is captured by this metric.
L_{day}	L_{day} is a descriptor of noise level based on the energy equivalent sound level ($L_{Aeq,T}$) over the daytime period, typically quantified over all day in a full year. Though the timings can vary from country to country, typical periods are 07:00 – 23:00 or 07:00 – 19:00 for countries which also use a separate evening metric, $L_{evening}$.

(continued)

(continued)

Term	Meaning
L_{den}	L_{den} is a descriptor of noise level based on energy equivalent sound level ($L_{Aeq,T}$) over a whole day ('den' refers to 'day, evening and night'). A penalty of 5 dB is applied to the evening period and 10 dB is applied to night-time period, recognising the increased sensitivity to sound during these periods.
$L_{evening}$	$L_{evening}$ is a descriptor of noise level based on the energy equivalent sound level ($L_{Aeq,T}$) over the evening period, typically quantified over all evenings in a full year. Though the timings can vary from country to country, a typical evening period is 19:00 – 23:00.
LFN	Low-frequency noise. Whilst the definition of 'low frequency' can vary, it is typically associated with sounds of 500 Hz or lower.
L_{Amax}	The maximum sound level identified during a measurement period. Experimental data has shown that the human ear does not generally register the full loudness of transient sound events of less than 125 ms duration and fast time weighting (F) has an exponential time constant of 125 ms which reflects the ear's response. Slow time weighting (S) has an exponential time constant of 1 s and is used to allow a more accurate estimation of the average sound level on a visual display. The maximum level measured with fast time weighting is denoted as $L_{Amax, F}$. The maximum level measured with slow time weighting is denoted $L_{Amax, S}$.
L_{night}	L_{night} is a descriptor of noise level based on the energy equivalent sound level ($L_{Aeq,T}$) over the night-time period, typically quantified over all nights in a full year. Though the timings can vary from country to country, a typical night-time period is 23:00 – 07:00.
Non-acoustic factors	Any parameter not relating to objective characteristics of a sound but that may influence response to sound, for example, socio-demographic parameters, attitudes towards the sound source, perception of fairness or control, etc.
Pa	Pascals. The unit of pressure used to determine Sound Pressure Level.
QALY	Quality Adjusted Life Year. A measure of health in which the benefits and length of life are adjusted to reflect quality of life. One QALY is equivalent to one year of full health.
SLM	Sound Level Meter. A specialist piece of equipment used for acoustic measurements.
Soundscape	An acoustic environment as perceived or experienced and/or understood by a person or people, in context.
Sound Power Level	The sound power level of a source is a measure of the total acoustic power radiated by a source. The sound power level is an intrinsic characteristic of a source (analogous to its volume or mass), which is not affected by the environment within which the source is located.

(continued)

(continued)

Term	Meaning
SPL	Sound Pressure Level. The sound power emitted by a source results in pressure fluctuations in the air, which are heard as sound and can be measured as a Sound Pressure Level. The Sound Pressure Level is ten times the logarithm of the ratio of the measured sound pressure (detected by a microphone) to the reference level of 2×10^{-5} Pa (the threshold of hearing).
Tranquillity	Tranquil areas are those that remain relatively undisturbed by noise and are prized for their recreational and amenity value for this reason.
WHO	The World Health Organization is a specialist agency of the United Nations responsible for international public health. The WHO has published several influential noise publications such as the Guidelines for Community Noise (1999), the Night Noise Guidelines for the European Region (2009) and the environmental Noise Guidelines for the European Region (2018).

Reference

The Chautauquan (1883), 3:9, p 543

Chapter 3
The Meaning(s) of Sound(s)

Rainer Guski

Abstract The meaning of sound for our lives and actions depends largely on the experience we have had with them, partly also on the context in which they occur. Sound events that humans and other animals can hear on earth are usually created by physical movement in the environment. *Meaning* implies that at least one (direct) relationship exists between the specific acoustic properties, and the perception of a sound, often associated with a second relationship: the perception of the sound and the identification of its source, e.g., a creaking door. And a third relation is usually associated, too: I like or dislike it (the sound and/or the source). Some relationships are rather trivial and are usually shared by a large cultural community. Other connections apply only in small groups of people (e.g., a neighborhood with a common history), and in extreme cases, only in a single person in a particular situation (e.g., a violin expert testing violins). This chapter starts with what dictionaries say about the word "sound" and the distinction between sound and noise. The next section deals with sounds as source of information. The last section describes the many non-acoustic influences on the meaning of sounds and noise, i.e., physical factors (e.g., color, object size, and motion), and personal factors (e.g., noise sensitivity, personal control, evaluation of the source, trust in the agents responsible, and expectations for the future acoustic development).

3.1 Introduction

'What's your mother tongue?' he had asked her. '*Português*,' she had answered.

The *o* she pronounced surprisingly as a *u*; the rising, strangely constrained lightness of the *é* and the soft *sh* at the end came together in a melody that sounded much longer than it really was, and that he could have listened to all day long (Mercier, Night train to Lisbon, 2008).

A chapter on *meaning* is indispensable in a book devoted to sound and noise. Sounds are not solid objects or even solid concepts that we can easily agree upon.

R. Guski (✉)
AG Environment and Cognition, Faculty Psychology, Ruhr-University, Bochum, Germany
e-mail: Rainer.Guski@ruhr-uni-bochum.de

© The Author(s) 2025
I. van Kamp and F. Woudenberg (eds.), *A Sound Approach to Noise and Health*,
Springer-AAS Acoustics Series, https://doi.org/10.1007/978-981-97-6121-0_3

33

The meaning of sound for our lives and actions depends largely on the experience we have had with them, partly also on the context in which they occur. For example, the quiet creaking of a car door outside our window can mean that the desired visitor is finally coming, or indicate that we still haven't oiled the door hinge, and/or that an unknown person is tampering with our car. M*eaning* implies that at least one (direct) relationship exists between the specific acoustic properties (e.g., the vibrations of a door in a certain frequency band for a certain time), and the perception of a sound, often associated with a second relationship: the perception of the sound and the identification of its source, a *creaking door*. And a third relation is usually associated, too: I like or dislike it (the sound and/or the source). Some relationships are rather trivial and are usually shared by a large cultural community. Other connections apply only to a small group of people (e.g. a neighborhood with a history of car theft), and in extreme cases, only to a single person in a particular situation (e.g. a collector of rare violins).

In addition, we should note the difference between semantic and pragmatic meanings. Semantic meaning is the literal meaning present in words. Pragmatic meaning is the meaning we add to the literal meaning, based on world knowledge or social communication habits. For example, the sound of an airplane belongs to the semantic class *sound*, but pragmatically it can rather be put into the semantic class *noise* if it is an unwanted sound. "The human relationship with sound is much deeper and more ancient than our relationship with words". (Kraus and Slater 2016, p. 84).

Each sound event that humans and other animals can hear on earth is created by physical movement in the environment, from the crackling of a small twig under our feet to vibrations of vocal cords or loudspeaker membranes. That is, except for rare situations in which we hear our own heart beating or biting into a dry pizza or eating a cracker, sounds that reach our ears usually indicate remote physical events, even beyond the boundaries of our visual field. In combination with the fact that our ears are always open—in contrast to the eyes which can be closed—the relationship between remote physical events and sound reaching the ear has led to the worldwide use of sounds for communication between animals, from short calls for food by young animals, to the rutting cries of adults, to the long love songs of blackbirds, nightingales and opera singers.

All animals, including humans, have learned to gather information from audible acoustic events; especially information about the source or cause of the event, its direction and relative distance from the ear, and in the case of the repetition of the sound, the speed of approach or removal. On the other hand, we are able to ignore sounds that we find unimportant, and relegate them to the background of our consciousness. Sometimes, we only become aware of a sound after it stops, like the humming of a refrigerator switching off. Learning to distinguish between sounds that are meaningful, and sounds that are not is a very essential aspect of hearing. In humans, hearing starts around week 22 in the womb. Of course, we don't know what the fetus experiences in the womb, but it seems that it often responds in an emotionally relaxed way to familiar relaxed voices and music from week 25 onward, while abrupt onset of loud sounds often is answered by baby startle movements.

This chapter on the meaning of sound starts with what dictionaries say about the word "sound" and the distinction between *sound* and *noise*. The next section deals with sounds as a source of information. The last section describes the many non-acoustic influences on the meaning of sounds.

3.2 Dictionaries on "Sound" and "Noise"

The Internet lexical entries on *sound* distinguish usually four different forms of the word "sound". For instance, the Britannica dictionary (https://www.britannica.com/dictionary/sound, seen 2024/01/18) distinguishes between the noun ("the sound"), the verb (e.g., "to cause (something) to make a sound or be heard", or "the buzzer sounded"), the adjective (e.g., "to be in a good condition"), and the adverb "sound" (e.g., "to be sound asleep"). Here, we will confine ourselves to the noun. However, it should be noted that the noun "sound" is often associated with a more or less vague linguistic modifier related to the perceived intensity (e.g., "loud", "soft") and/or spectral composition (e.g., "harsh", "dark", "bright").

The Britannica dictionary of SOUND as a noun shows several meanings:

1: something that is heard [or can be heard, (R.G.)].

Examples: the *sound* of footsteps/thunder / the *sounds* of laughter / I heard a loud, buzzing *sound*/I didn't hear a *sound*. [=I didn't hear anything]/They never made a *sound*/speech *sounds* [=the sounds people make when they speak words]/devices used to record *sound* ...

2: the speech, music, etc., that is heard as part of a broadcast, film, or recording.

Examples: The film was good, but the *sound* was poor. [This relates to the acoustic quality of the soundtrack; (R.G.)]/digital *sound [This may relate to a somewhat harsher sound quality of early digital recorders, as compared to analogue recorders, it may also relate to early attempts of Digital Audio Workstations to imitate musical instruments; (R.G.)]/*"Can you turn up the *sound*? [=volume] I can't hear what they're saying."

3: the particular musical style of an individual, a group, or an area—usually singular.

Examples: the Nashville *sound*/"I like the band's *sound*".

4: the idea that is suggested when something is said or described.

Examples: "The doctor says my case is unusual.—I don't like the *sound* of that." [=I don't like the way that sounds; that sounds bad/serious]/By/from the *sound* of it, you may have poison ivy./*sound* and fury: loud and angry words that attract a lot of attention but do nothing useful.

Personally, I like to add some other meanings of the word *sound* to those in the Britannica, e.g.,

- the *sound* of rooms, e.g. concert halls, tunnels, libraries and bathhouses. This refers mostly to the acoustic quality of the room in terms of reverberation time due to the sound reflecting surfaces and typical behavior of people inside such rooms

- the *sound* of public places. In this case, we refer to the acoustic content of these places, e.g., traffic, people, reflecting walls, etc. (see *soundscape*)
- the *sound* of a certain language. Even if we don't understand the exact meaning of spoken words, we often are able to distinguish language families, like Romance (e.g., French, Spanish, Italian), Germanic (e.g., English, Dutch, German), Slavic (e.g., Polish, Czech, Slovak) and some Asian languages due to the prosody, pitch, pronunciation of accents, speed and phonology
- the *sound* of children's/female/male voice: voices of young boys and girls are very similar before they hit puberty. Their fundamental frequency and formant frequencies are higher in pitch as compared to adults, and such variables can be used to identify children's speech (Barreda and Assmann 2021). With increasing age, male and female voices usually differ in fundamental frequency and formant frequency (female being higher), as well as a breathier voice quality is sometimes reported with females (e.g., Titze 1989; Weirich and Simpson 2018).

Other languages than English have their own meanings of the equivalents to "sound". For instance in German, *sound* is associated with four nouns: *Klang, Geräusch, Ton* and *Schall. Klang* is associated with timbre, i.e., the energy distribution over the frequency spectrum of the sound, as well as reverberation. It is often used in the context of music and room acoustics. *Geräusch* is reminiscent of *Rauschen* (related to *swoosh* and *hissing* in English) and is often used when the source of the sound is unclear. *Ton* mostly relates to a sound event with a rather small frequency band, as used in single-tone music (e.g., vocal solo, ringtones), as well as in some electronic alarm systems. Another common meaning of *Ton* relates to the manner, or tone of speaking. *Schall* is very broad and rather neutral, it may relate to every sound and does not imply any perceptual evaluation.

Some word combinations with "sound" are of relevance to this book on sound and noise:

- *Sound art* produces sound installations, mostly in the context of public events, exhibitions, films and concerts. It goes back to futurist movements at the beginning of the twentieth century that responded to the increase in the spread of mechanical and industrial sounds (Rudi 2008). Due to the rapid development of electronics after World War II, it became increasingly possible to alienate everyday sounds and create new sounds. Sounds could now be detached from their mechanical source and thus treated abstractly as "pure ideas" (Russolo 1913). This led on the one hand to new forms of music, on the other hand, to sound installations without musical components (e.g., → *sound sculptures*), and furthermore to the possibility of changing the acoustic properties of a room with software and loudspeakers and is not limited to the design of the room walls. Also, see Chap. 9 by Marcel Cobussen.
- *Sound design* or *sound-design*, the creative work on all acoustic elements of a presentation or a product with the exception of music, i.e. dialogue, sounds, atmospheres and sound effects. Depending on the budget and intended use, these sound

elements are produced and recorded and/or taken from sound archives and then edited. Early examples of sound design can be found in product design, e.g., car engines and car doors, or vacuum cleaners and food processors.

- *Soundscape*, e.g., the sound of cities, places, pubs and restaurants. "Soundscape suggests exploring all of the sound in an environment in its complexity, ambivalence, meaning, and context. Basically, the soundscape concept considers the conditions and purposes of its production and perception. Consequently, it is necessary to understand that the evaluation of noise/sound is a holistic approach" (Brooks et al. 2014, p. 30).
- *Soundwalk*, a guided walk through a certain environment, can take many forms and methods. "Independent of the method applied, the ultimate aim of sound walking is to listen consciously to the environment and to increase our awareness of the quality of the sonic environment" (Radicchi 2017, p. 73) (Fig. 3.1).

The Britannica dictionary seems to equate *loud sounds* with *noise*. It has no special entry on *noise*; however, it provides an entry for *noise pollution*. The Cambridge Dictionary defines noise as "a sound or sounds, especially when it is unwanted,

Fig. 3.1 Gino Severini, 1912, *Dynamic Hieroglyphic of the Bal Tabarin*, oil on canvas with sequins, 161.6 × 156.2 cm (63.6 × 61.5 in.), Museum of Modern Art, New York. There are many incongruent movements at the same time – they will cause many incongruent sounds

unpleasant or loud". It provides several examples of environmental noise, e.g., street traffic noise. It also provides specialized meanings for noise, like electronic noise, and statistical noise. In the end, it declares *sound* and *noise* both "refer to something which you can hear, but when a *sound* is unwanted or unpleasant, we call it a *noise*" (https://dictionary.cambridge.org/dictionary/english/noise, seen 2024/01/18).

In the realm of noise effects research, the latter meaning of the term *noise* is generally accepted: when a *sound* is unwanted or unpleasant, we call it noise (e.g., Goldsmith and Jonsson 1973; Guski 1976; Stallen 1999). However, who decides about the undesirability? In the field of environmental noise, it is usually the 'receiver' who evaluates the sound as noise, even if the 'sender' evaluates the sound positively. From this point of view, *noise* is a psychological term. On the other hand, sounds with high volume can harm the health of people affected even if they find the sound acceptable, as in the case of deafening party music. So psychology is not enough here; other findings, e.g. from medicine, must be added to protect the population from the consequences of loud sounds.

3.3   Sound Events as Source of Information

Sight and hearing are among the most important abilities of human beings—we usually can use both of them very well, but they obviously serve different purposes. This is shown, for example, by the fact that our hearing is open to acoustic information 24 h a day—we cannot close our ears without technical aids, while our eyes are closed for an average of 7 h a day. This means, among other things, that hearing has a distinct advantage over vision in monitoring tasks: We can, for instance, even in complete darkness, detect that someone is running behind us, with increasing speed and decreasing distance from us. The acoustic events we perceive are based on mechanical events, predominantly movements of certain objects (masses) relative to other objects. The soles of our pursuer's shoes on the ground produce certain frictional sounds for example. The loudness, frequency spectrum, and onset and decay time of the friction sounds provide information about the material properties of the bodies involved in the friction, e.g., the volume and material of the shoe sole and the roughness of the floor. In addition, the repetition rate of the successive step sounds, together with the change in volume over time, provide information about the spatial proximity of the pursuer.

Of course, the relations between acoustic events and material properties of bodies (masses) have to be learned, similar to the way the relations between texture, color, brightness gradient and edges have to be learned as information about objects or surfaces in vision. However, learning occurs differently in the course of child development in vision and hearing. In vision, we usually deal with persistent objects that remain existent in our perception even if we move relative to them or briefly close our eyes. With hearing we are dealing with non-persistent, i.e. transient, events. In order to learn the object properties, we have to make sure that the acoustic events are repeated—usually, we (and even more so our children) use collisions for this purpose: we tap several times on the objects and change the tapping force in the process, sometimes also the position of the collision point and possibly the object with which we knock.

While learning acoustic object properties by repetition, we also learn to tolerate variations. For instance, a child learns that the mother's voice may change in volume, pitch, and tempo to a certain degree, and yet is the voice of the same person. Tolerance or intolerance of variation has been and is often studied in the field of music. This did not always turn up consistent results, but we can note that people with only a low level of musical education, for example, can rarely distinguish the transformation of a melody from major to minor, but can recognize that it is basically a very similar melody that differs from other melodies tested (Bigand and Poulin-Charronat 2006; Daikoku 2018).

We usually do not need any training in order to localise a sound source—at least in uncluttered environments and, after some training, we can easily name the sound source. We usually *experience* sounds in terms of their sources—we hear ravens croaking loudly, cars driving fast, children playing loudly—and not the sounds themselves (Gaver 1988, p. 1).

We as humans are apt in identifying sounds with often a clear understanding of the sources that produce them. However, the evaluative part of the meaning may change depending on acoustic, situational, and temporal circumstances. For instance, so-called approaching (or "looming") sounds, i.e., sounds with increasing amplitude over a certain time are generally more annoying than so-called "receding" sounds, i.e., sounds with decreasing amplitude (Rosinger et al. 1970; Bach et al. 2009). On the other hand, approaching footsteps heard in our living room may be welcome when heard as coming from outside at a time when we expect guests. The same sound will take on a much less positive meaning if it occurs at night, or appears to be coming from the apartment above us, or at a higher volume, or at a higher walking speed (Frescura et al. 2022).

Intrusive and alarming sounds are of specific relevance to us. Intrusiveness may explain why some low-level sounds are annoying, for instance, heel clicks in apartment buildings or indistinct conversation. According to Fidell and Teffeteller (1978) low-level sounds intrude upon awareness if they are highly detectable in the background mixture of sounds. Anna Preis (1987) assumed that the intrusive character of such sounds is caused by their roughness (rapid amplitude modulation) or dissonance (harsh, unpleasant sound).

Alarming sounds are highly relevant. A newborn's first scream is probably the only one that is truly welcomed by the parents. This first cry is probably more of a cry of distress for the baby, because it urgently needs oxygen and must now supply the lungs, which were previously supplied by the placenta. Nevertheless, this cry is welcomed by adults because it signals a new and independent life. Other cries are often considered more like alarm signals. This is due to the pronounced acoustic roughness. Not all cries are signals that warn of danger or indicate pain. Frühholz et al. (2021) were able to differentiate 6 distinct types of human screams and found that cries of joy accelerate cognitive processes, whereas cries of distress or alarm tend to slow them down, in some cases even paralyzing them.

We know many other acoustic alarm signals, often it is our own cell phone, the kitchen alarm clock, the fire engines, police, horns, brake squeals, the front doorbell, and many more. Most alarm signals in our everyday environment have a tonal structure and begin with an immediate rise in level, followed by a tonal change or pause. Depending on the urgency and cause, this sequence is repeated. In the clinical context, many different auditory alarms are used, some of which sound simultaneously—which has sometimes led to confusion, 'false alarms' and 'alarm fatigue' among staff (McDougall et al. 2020).

So far, we handled single sound events as information about separate environmental events. At a first glance, this seems inadequate for our everyday situation where different sound events follow each other in time, or may even overlap in time. For instance, at the time of writing these sentences, I am sitting in my university office, a window is open, and there is a continuous grinding noise with intermittent tonal episodes (this seems to come from a large construction site on the university campus). Every now and then, I hear a voice in the hallway outside my office, then a door slamming, and just now the motorized sunshade in the neighboring office is being lowered. The fact that all these sounds partly overlap and yet can be perceived separately is due to *auditory stream segregation* (Bregman 1990). "A sequence of sounds may be heard as coming from a single source (called fusion or coherence) or from two or more sources (called fission or stream segregation). Each perceived source is called a 'stream'. When the acoustic differences (or time delays) between successive sounds are very large, fission nearly always occurs, whereas when the differences are very small, fusion nearly always occurs. When the differences are intermediate in size, the percept often 'flips' between one stream and multiple streams, a property called 'bistability'. The flips do not generally occur regularly in time. The tendency to hear two streams builds up over time, but can be partially or completely reset by a sudden change in the properties of the sequence or by switches in attention" (Moore and Gockel 2012, p. 919).

3.4 Non-Acoustic Influences

Sometimes, children maintain that red cars are more powerful and louder than green cars. Most adults will reject this statement and at best reply, "The cars themselves are probably equal from the technical side, but it may be that drivers of red cars like to drive in a more powerful manner than drivers of green cars." Of course, this would be a wild assumption, and an empirical test very difficult to undertake. Anyway, the idea that the color of a sound source may have an influence on its perceived loudness is a simple example of a possible non-acoustic factor (color) on the perception of sounds.

The scientific study of the contribution of non-acoustic factors in the assessment of auditory events has turned out to be one of the most interesting, albeit complex, undertakings of modern times; the European Union has funded and continues to fund projects in which non-acoustic factors are involved (e.g., https://anima-project.eu/ and htttp://famos-study.eu/).

In a future ISO standard non-acoustic factors are defined as: '*All factors other than the objective, measured or modelled acoustic parameters which influence the process of perceiving, experiencing and/or understanding an acoustic environment in context, without being part of the causal chain of this process*' (Fenech et al. 2021, p. 4). Non-acoustic factors can be grouped into *physical factors* (e.g., color, size, and motion), *personal factors* (e.g., noise sensitivity and attitudes toward the sound source), *social factors* (e.g., socially shared experiences with a local noise source), *situational factors* (e.g., time of day, access to a quiet façade), and *contextual factors* (e.g., infrastructural change situations). These factors partly overlap. For example, personal experiences are in part also socially shared experiences, and the presence of a quiet facade is not only a situational factor, but also an acoustic one. For a detailed review, see (Riedel et al. 2021).

3.5 Physical Factors

I like to start here with some basic scientific questions about auditory sound perception of events and their relation to non-acoustic factors like visible properties and language use. The importance of the latter will become clear when we consider real-life events, e.g., the passing-by of a "sports car" or a lorry. Both events can be very loud, but even in the case of constant velocity the sounds will at least differ in spectral content and sharpness (the lorry will sound deeper due to a slower rate of ignitions and larger tires, and it will sound sharper due to the type of ignitions in case of Diesel engines). The linguistic classification of the two sounds in terms of sound frequency (lower vs. higher) will probably conjoin the semantic parameter of *size*, correlating high-frequency sounds with small size and low-frequency sounds with large size (Spence 2011; Johansson et al. 2020).

Color: With respect to the influence of *color on loudness*, there have been several congress papers (e.g. Fastl 2004), and journal contributions discussing outcomes of studies on this topic (e.g. Menzel et al. 2008). In sum, the effect of a vehicle color (mainly hue and luminance) on vehicle loudness is generally very small, when experimental conditions are carefully controlled for. And there are many controls necessary: The scientific analysis of the relations between color on sound must at least take care to control for basic color properties such as luminance, saturation, and hue as well as basic acoustic properties such as loudness, pitch, and spectral composition (Anikin and Johansson 2019), and in case of near-realistic events control for direction and speed of movements, as well as the corresponding effects of light and sound reflections.

Object size and loudness: In contrast to object color, object size seems less arbitrarily related to loudness. The relation between size and loudness can be seen as an example of prothetic cross-modal correspondences—a prothetic correspondence is based on the amplitude or amount rather than on the quality of sensory experience, e.g., "large loudness" = "large size". Unfortunately, the (perceived) loudness of a sound does not only depend on the size of the source pitch and (partially) duration as well as the velocity of motion also plays a significant role. Houben (Houben 2002, Exp. 1) used recordings of wooden balls of different sizes and speeds rolling over a wooden plate as stimuli. The participants were asked to compare two sounds and indicate the larger of the balls heard. The results show that participants reliably indicated the larger ball, as long as the difference between the size of the two balls exceeded 14% in diameter. However, in this example, the size of the balls may be causally related to the acoustic features of the rolling balls—this would contradict the tentative definition of "non-acoustics factors" given by Fenech et al. (2021). On the other hand, if the prothetic relationship works, humans would at least assume that small objects in a certain class of sound-emitting objects (e.g., cars, mechanical tools, or music instruments) produce softer sounds than large objects. Seen this way, reducing the size of certain loud objects may contribute to a reduction of noise annoyance.

Motion: Most of the environmental sound sources in the vicinity of residential buildings move, i.e., they change their location in a certain direction and with a certain speed in relation to the receivers. On the other hand, there are stationary sound sources, such as factories, craft workshops, electrical substations, wind turbines, ventilation systems and heat pumps. Considering that the forms of movement of road, rail and air traffic are not included in the usual calculations of the average levels for the means of transport, just as possible side effects of industrial noises (e.g. odors), a comparison of the effects of these sources at comparable sound levels seems daring, if not questionable. Nevertheless, it is surprising that the exposure–response curve (ERF) published by Janssen et al. (2009) for percentages of persons highly annoyed by (stationary) industrial noise (% HA) in the range between 45 and 60 dB L_{den} is only slightly below the ERF for (moving) air traffic - and significantly above the ERF for road traffic according to Miedema and Oudshoorn (2001). Does this mean that the annoyance caused by (stationary) industrial noise in the vicinity of residential areas is almost similar to the annoyance caused by (moving) aircraft noise at computationally comparable sound levels? I am afraid that if we compare the two, we forget all too quickly that industrial plants have other effects besides noise—for example, they are usually highly visible during the day and at night and shape entire neighborhoods.

3.6 Personal Factors

A lot of research into non-acoustic variables is done on noise annoyance often in relation to traffic noise. The major factors having a large influence on noise annoyance are individual noise sensitivity, degree of personal control over noise, evaluation of the noise source, trust/distrust in the agents responsible for noise, and expectations for the future of noise development.

3.6.1 Noise Sensitivity

Noise sensitivity is one of the most influential factors in noise annoyance. 10 to 15% of the population is highly noise sensitive. Researchers differ somewhat in how they define noise sensitivity. Stansfeld (Stansfeld 1992) sees noise sensitivity as more or less stable and related to critical attitudes towards the environment: "Noise sensitive people attend more to noises, discriminate more between noises, find noises more threatening and out of their control, and react to, and adapt to noises more slowly than less noise sensitive people". And, after discussing psychiatric symptoms as consequences of environmental stressors: "The *meaning* of the noise for the individual is the central feature of such vulnerability to noise effects, typified by noise sensitivity" (Stansfeld 1992, p. 39, Italics: R.G.). Job (1999) conceptualized noise sensitivity as internal states: one related to loud noises (road traffic, lawn mower), and the other related to quieter noise situations which are nonetheless distracting (rustling papers at the movies, people talking while watching television)". Ellermeier et al. (2001) concluded that noise-sensitive people do not have better sensory capacities, but differ with respect to the evaluation of sounds. Schütte et al. (2007) saw indications that noise sensitivity differs between domains, e.g., leisure, work, habitation, communication, and sleep. Therefore, they developed the Noise-Sensitivity-Questionnaire (NoiSeQ) in order to measure global noise sensitivity as well as the sensitivity to the five domains mentioned above.

Shepherd et al. (2015) see noise sensitivity as a psychological trait, i.e., as a stable personality attribute describing the degree of tolerance to sound. The authors did not find support for the notion that, by itself, negative affectivity explains noise sensitivity. Welch et al. (2022) found evidence for three different aspects of noise sensitivity depending on the individual meaning of the noise (Fig. 3.2).

3.6.2 Personal Control

Except, to a certain extent, for neighborhood noise, residents of noisy areas usually have little personal control over the noise emitting source (e.g., road, rail, and air traffic). In case of neighborhood noise, they may try to talk to the neighbor and negotiate a lower volume or quiet time. However, the more or less anonymous sources of environmental noise are controlled by complex laws and regulations over which individual affected citizens have very little influence. They can join forces with neighbors and protest to the authorities, but success is hard to come by—usually, the result is no more than passive noise protection, e.g. soundproof windows that are best kept closed at all times. Such experiences or expectations contribute to the fact that those affected feel more or less at the mercy of the noise - especially when those affected cannot change their place of residence. It is well known that low degrees of personal control over noise contribute to increased noise annoyance,

Fig. 3.2 "Lauschende Frau" (Listening woman) by Andreas Schnelle, 1995. Oil on canvas. ©with permission. The lady seems to be very sensitive to sounds

diminished quality of life, and partially to increased health risks (e.g., Hatfield et al. 2002; Bartels et al. 2022).

3.6.3 *Evaluation of the Noise Source*

Negative evaluation of sound is implied in the word noise – we don't like this certain sound. And, generally, we don't like the ***source*** of the noise. Some sources of noise we particularly dislike, e.g., because they endanger us physically (e.g., flying machines can crash down on us), others we dislike because they are active at night, others we dislike because they additionally pollute the air, affect the value of our home and/ or because they do not make sufficient efforts to avoid or reduce noise. In case of man-made sounds, exposing somebody to sounds or being exposed to sounds is a form of social interaction ('You expose me'; Stallen 1999; Maris 2008) in which the evaluation of the source and aspects of fairness play an important role. In some cases, many factors come together at the same time. Much research on the effects of source evaluation on noise annoyance and reported sleep disturbance has been done with respect to aircraft noise. This is understandable because it has been known for some time that aircraft noise is generally more annoying than road and rail noise at comparable continuous sound levels (Miedema and Oudshoorn 2001; Guski et al. 2017).

An early meta-analysis of Fields (1993), based on 136 field studies showed... "firm evidence that noise annoyance is associated with: the fear of an aircraft crashing or of danger from nearby surface transportation; the belief that aircraft noise could be prevented or reduced by designers, pilots, or authorities related to airlines; and an expressed sensitivity to noise generally (not only local environmental noise)" (Fields 1993, p. 2757 f.). The item "Fear of Aircraft Crashing in the Neighborhood" has for years been a common part of questionnaire studies in the vicinity of airports, particularly in those countries that have air traffic with a relatively high proportion of private and smaller commercial aircraft, or an aircraft crash in the past.

3.6.4   Trust in the Agents Responsible for Noise

Research (Job 1988; Schreckenberg et al.2017) finds a clear and often strong relationship between the degree of trust in authorities responsible for the operation of a noise source and annoyance. However, it is unclear what the causal relationship is. The above-mentioned studies both found that trust determines annoyance as well as the reverse: annoyance determines trust. There seems to be a complex pattern in the relation between trust in authorities and noise annoyance which also changes over time. A conclusion that several authors draw from this complex pattern is that there is a need to increase the 'trust' policy of the airport (see also Kroesen and Bröer 2009) and "to engage airport communities in aviation-related decision making by improving the information and communication of airports in order to enhance residents' 'competence' and also trust in the airport noise authorities" (Heyes et al. 2022, p. 219).

3.6.5   Expectations for the Future of Noise Development

It seems self-evident that residents' expectations have a great influence on annoyance judgments, especially in those situations where local infrastructure changes—e.g., when roadways, railways, or airways in the vicinity are expanded or deconstructed. The hypothesis is: if residents expect their noise exposure to increase in the future, they will already be more annoyed beforehand than if they expect constant exposure. And if residents expect their noise exposure to be lower in the future, they will already be less annoyed beforehand than if constant exposure is expected. In other words, expectation changes annoyance in the direction of expectation.

A drastic example of the expectation effect is reported in Mense and Kholodilin (2014). The authors studied internet house prices in the vicinity of the Berlin International Airport BER (the former Schönefeld Airport) during the airport planning process from April 2011 to December 2012. The planned flight paths were published in July 2011 and (revised) in January 2012. It turned out that property listing prices were reduced substantially (9.6% on average) in the affected areas after the flight paths were published.

In the longitudinal study at Frankfurt Airport already mentioned above (Schreckenberg et al. 2017), the authority asked residents about (positive) expectations concerning the impact of air traffic on regional development and residential quality of life. An increase of 1 point on a 5-point expectation scale (from 1 to 2 and from 3 to 4 more positive) resulted in a decrease of highly annoyed residents with respectively 24 and 31%. Such results should encourage operators of noisy infrastructures to credibly present the benefits of their noise source, if it cannot be operated more quietly.

3.7   Conclusions

This chapter explores relations between sounds and certain human cognitions: M*eaning of Sound* implies that at least one (direct) relationship exists between specific acoustic properties of an audible event (e.g., the rise and decay of vibrations in a certain frequency band for a certain time), and the perception of a sound, often associated with a second relationship: between the perception of the sound and the identification of its source. And a third relation is usually associated, too: I like or dislike it (the sound and/or the source). Sounds that reach our ears usually indicate remote physical events, even beyond the boundaries of our visual field. Since our ears are open 24 h and can neither be closed at will nor effectively focused in a way that "isolates" a certain sound event in the multitude of sounds surrounding us at the same time, we need cognitive processes in order to follow a certain sound source. Such cognitive "streaming" processes can be trained, however, they are rather helpless in case of environmental sounds much louder than the sounds we want to follow. The situation may be welcome in case of alarm signals which inform about impending danger. However, in the majority of loud environmental sounds, we evaluate them as noise, i.e., unwanted sounds. There is much research about the acoustic properties of unwanted sounds and their effects on people (not covered in this chapter). This chapter lists and discusses a series of non-acoustic physical and personal factors contributing to the evaluation of sounds and noise situations. These factors may partly help to understand the wide variety of reactions of residents in noisy situations.

References

Anikin A, Johansson N (2019) Implicit associations between individual properties of color and sound. Atten Percept Psychophys 81(3):764–777. https://doi.org/10.3758/s13414-018-01639-7

Bach D, Neuhoff JG, Perrig W, Seifritz E (2009) Looming sounds as warning signals: The function of motion cues. Int J Psychophysiol 74:28–33. https://doi.org/10.1016/j.ijpsycho.2009.06.004

Barreda S, Assmann PF (2021) Perception of gender in children's voices. J Acoust Soc Am, 150(3949). https://doi.org/10.1121/10.0006785

Bartels S, Richard I, Ohlenforst B, Jeram S, Kuhlmann J, Benz S, Schreckenberg D (2022) Coping with Aviation Noise: Non-Acoustic Factors Influencing Annoyance and Sleep Disturbance from Noise. In Leylekian L, Covrig A, Maximova A (Eds), Aviat Noise Impact Manag (pp. 197–218): Springer Open Access. https://doi.org/10.1007/978-3-030-91194-2_8

Bigand E, Poulin-Charronnat B (2006) Are we '"experienced listeners"'? A review of the musical capacities that do not depend on formal musical training. Cognition 100:100–130. https://doi.org/10.1016/j.cognition.2005.11.007

Bregman AS (1990) Auditory scene analysis. MIT Press, Cambridge, USA, MA

Brooks BM, Schulte-Fortkamp B, Voigt KS, Case AU (2014) Exploring our sonic environment through soundscape. research & theory. How can we know what people think of their sonic environment? Well, we ask them! Acoustics Today, Winter 2014, 30–40. Retrieved from https://acousticstoday.org/wp-content/uploads/2015/05/Exploring-Our-Sonic-Environment-Through-Soundscape-Research-Theory.pdf (2022/07/25)

Daikoku T (2018) Time-course variation of statistics embedded in music: Corpus study on implicit learning and knowledge. Plos One, 13(e0196493). https://doi.org/10.1371/journal.pone.0196493

Ellermeier W, Eigenstetter M, Zimmer K (2001) Psychoacoustic correlates of individual noise sensitivity. J Acoust Soc Am 109:1464–1473. https://doi.org/10.1121/1.1350402

Fastl H (2004) Audio-visual interactions in loudness evaluation. Paper presented at the 18th International Congress on Acoustics, Kyoto, Japan

Fenech B, Lavia L, Rodgers G, Notley H (2021) Development of a new ISO Technical Specification on non-acoustic factors to improve the interpretation of annoyance and soundscape datasets. Paper presented at the ICBEN 2021, Stockholm, Sweden

Fidell S, Teffeteller S (1978) The relationship between annoyance and detectability of low level sounds. DC (USA), BBN Report, Washington, p 3699

Fields JM (1993) Effect of personal and situational variables on noise annoyance in residential areas. J Acoust Soc Am 93:2753–2763

Frescura A, Lee PJ, Soeta Y, Ariki A (2022) Effects of spatial characteristics of footsteps sounds and non-acoustic factors on annoyance in lightweight timber buildings. Build Environ 222:109405. https://doi.org/10.1016/j.buildenv.2022.109405

Frühholz S, Dietziker J, Staib M, Tros W (2021) Neurocognitive processing efficiency for discriminating human non-alarm rather than alarm scream calls. PLoS Biol 19(4):e3000751. https://doi.org/10.1371/journal.pbio.3000751

Gaver WH (1988) Everyday listening and auditory Icons. (Dissertation). University of California, San Diego, San Diego, CAL, USA. Retrieved from ResearchGate 2022/07/21

Goldsmith JR, Jonsson E (1973) Health effects of community noise. Am J Public Health 63(9):782–793. https://doi.org/10.2105/ajph.63.9.782

Guski R (1976) Der Begriff 'Lärm' in der Lärmforschung [The notion of noise in noise research]. Kampf dem Lärm 23:43–52

Guski R, Schreckenberg D, Schuemer R (2017) Review: WHO environmental noise guidelines for the european region: a systematic review on environmental noise and annoyance. Int. J. Env. Res. Pub. Health 14(12):1–41. https://doi.org/10.3390/ijerph14121539

Hatfield J, Job RFS, Hede AJ, Carter NL, Peploe P, Taylor R, Morrell S (2002) Human response to environmental noise: the role of perceived control. Int J Behav Med 9(4):341–359

Heyes G, Hauptvogel D, Benz S, Schreckenberg D, Hooper P, Aalmoes R (2022) Engaging Communities in the Hard Quest for Consensus. In Leylekian L, Covrig A, Maximova A (Eds). Aviat Noise Impact Manag, (pp. 219–239). Cham: Springer Open Access, https://doi.org/10.1007/978-3-030-91194-2_9

Houben M (2002) The sound of rolling objects. Perception of size and speed. (PhD Thesis). Tech Univ Eindh, Eindh, NL. https://doi.org/10.6100/IR556897

Janssen SA, Vos H, Eisses AR, Pedersen E (2009) Exposure-response relationships for annoyance by wind turbine noise: a comparison with other stationary sources. Paper presented at the 8th European Conference on Noise Control 2009 (EURONOISE 2009), Edinburgh, UK

Job RFS (1988) Community response to noise: A review of factors influencing the relationship between noise exposure and reaction. J Acoust Soc Am 83:991–1001

Job RFS (1999) Noise sensitivity as a factor influencing human reaction to noise. Noise Health 1(3):57–68

Johansson NE, Anikin A, Carling G, Holmer A (2020) The typology of sound symbolism: Defining macro-concepts via their semantic and phonetic features. Linguist Typology, 24(2). https://doi.org/10.1515/lingty-2020-2034

Kraus N, Slater J (2016) Beyond Words: How Humans Communicate Through Sound. Annu Rev Psychol 67(1):83–103. https://doi.org/10.1146/annurev-psych-122414-033318

Kroesen M, Bröer C (2009) Policy discourse, people's internal frames, and declared aircraft noise annoyance: An application of Q-methodology. J Acoust Soc Am 126:195–207. https://doi.org/10.1121/1.3139904

Maris E (2008) The social side of noise annoyance. (PhD thesis). University of Leiden, Leiden (NL). https://openaccess.leidenuniv.nl/bitstream/handle/1887/13361/Thesis_Maris_A4.pdf?sequence=3. Last access: 2023/02/08

McDougall S, Edworthy J, Sinimeri D, Goodliffe J, Bradley D, Foster J (2020) Searching for meaning in sound: Learning and interpreting alarm signals in visual environments. J Exp Psychol Appl 26(1):89–107. https://doi.org/10.1037/xap0000238

Mense A, Kholodilin KA (2014) Noise expectations and house prices: the reaction of property prices to an airport expansion. Ann Reg Sci 52(3):763–797. https://doi.org/10.1007/s00168-014-06

Menzel D, Fastl H, Graf R, Hellbrück J (2008) Influence of vehicle color on loudness judgments. J Acoust Soc Am 123:2477–2479. https://doi.org/10.1121/1.2890747

Miedema HME, Oudshoorn CG (2001) Annoyance from transportation noise: Relationships with exposure Metrics DNL and DENL and their confidence intervals. Environ Health Perspect 109:409–416. https://doi.org/10.1289/ehp.01109409

Moore BC, Gockel HE (2012) Properties of auditory stream formation. Philos Trans R Soc Lond B Biol Sci 367(1591):919–931. https://doi.org/10.1098/rstb.2011.0355

Pascal Mercier, Night train to Lisbon, 2008

Preis A (1987) Intrusive sounds. Appl Acoust 20(2):101–127. https://doi.org/10.1016/0003-682X(87)90057-0

Radicchi A (2017) A Pocket Guide to Soundwalking. Some introductory notes on its origin, established methods and four experimental variations. In Besecke A, Meier J, Pätzold R, Thomaier S (Eds.), Stadtökonomie—Blickwinkel und Perspektiven, Ein Gemischtwarenladen. Perspectives on Urban Economics A General Merchandise Store. Berlin, Germany: Universitätsverlag Technische Universität Berlin

Riedel N, Kamp IV, Dreger S, Bolte G, Andringa T, Payne SR, Paviotti M (2021) Considering 'non-acoustic factors' as social and environmental determinants of health equity and environmental justice. Reflections on research and fields of action towards a vision for environmental noise policies Transportation Research Interdisciplinary Perspectives, 11(100445). https://doi.org/10.1016/j.trip.2021.100445

Rosinger G, Nixon CW, von Gierke HE (1970) Quantification of the noisiness of 'approaching' and 'receding' sounds. J Acoust Soc Am 48(4):843–853

Rudi J (2008) Sound and Meaning. In J. Rudi (Ed.), Absorption and resonance—sound and meaning (pp. 118–127). Oslo, Norway: NOTAM. Retrieved from ResearchGate 2022/07/21/

Russolo, Luigi (1967) The art of noise:(futurist manifesto, 1913). Primary Information

Schreckenberg D, Benz S, Kuhlmann J, Conrady M, Felscher-Suhr U (2017) Attitudes towards authorities and aircraft noise annoyance. Sensitivity analyses on the relationship between non-acoustical factors and annoyance. Paper presented at the ICBEN 2017, Zurich, Switzerland

Schütte M, Marks A, Wenning E, Griefahn B (2007) The development of the noise sensitivity questionnaire. Noise Health 9:15–24

Shepherd D, Heinonen-Guzejev M, Hautus M, Heikkilä K (2015) Elucidating the relationship between noise sensitivity and personality. Noise Health 17(76):165–171. https://doi.org/10.4103/1463-1741.155850

Spence C (2011) Crossmodal correspondences: A tutorial review. Atten Percept Psychophys 73:971–995. https://doi.org/10.3758/s13414-010-0073-7

Stallen PJ (1999) A theoretical framework for environmental noise annoyance. Noise Health 1(3):69–80

Stansfeld SA (1992) Noise, noise sensitivity and psychiatric disorder: epidemiological and psychophysiological studies. Psychol Med Monogr 22:1–44

Titze IR (1989) Physiologic and acoustic differences between male and female voices. J Acoust Soc Am, 85(4). https://doi.org/10.1121/1.397959

Weirich M, Simpson AP (2018) Gender identity is indexed and perceived in speech. Plos One, 13(12). https://doi.org/10.1371/journal.pone.0209226

Welch D, Dirks KN, Shepherd D, Ong J (2022) What is Noise Sensitivity? Noise Health 24(114):158–165. https://doi.org/10.4103/nah.nah_56_21

Chapter 4
Noise and Effects on Health and Well-Being

Charlotte Clark, Danielle Vienneau, and Gunn Marit Aasvang

"Sound is a silent killer. The WHO has calculated that in West-Europe alone, every year, 1 million healthy life years are lost due to sound pollution".

Quote by Kirsten van den Bosch (University of Groningen, The Netherlands) Sound is a silent killer Studium Generale University of Utrecht, The Netherlands, 2019.

This chapter summarizes the most recent evidence on how unwanted or harmful sounds, primarily from transportation, can affect health.

4.1 What is Noise? Noise Indicators, Environmental Noise Directive

Noise is often defined as "unwanted and/or harmful sound" (Fink 2019). Very loud impulse sounds or prolonged exposure to high sound levels can damage hearing. Sound or noise from traffic, industry and other commercial activities in residential areas does not reach levels that are harmful to hearing but can cause disturbances in activities such as communication, rest and sleep, and be perceived as annoying. Noise can act as a non-specific stressor both during the day and at night. There are large individual differences in how we experience noise, and the effects are dependent on

C. Clark
St George's, University of London, London, UK
e-mail: chclark@sgul.ac.uk

D. Vienneau (✉)
Swiss Tropical and Public Health Institute, Allschwil, Switzerland
e-mail: danielle.vienneau@swisstph.ch

G. M. Aasvang
Department of Air Quality and Noise, Norwegian Institute of Public Health, Oslo, Norway
e-mail: GunnMarit.Aasvang@fhi.no

© The Author(s) 2025
I. van Kamp and F. Woudenberg (eds.), *A Sound Approach to Noise and Health*, Springer-AAS Acoustics Series, https://doi.org/10.1007/978-981-97-6121-0_4

acoustic factors as well as the conditions of the situation and the person experiencing the noise.

Sounds from most of the sources that we are exposed to in our surrounding environments vary in intensity over time. So, how can we best describe the noise that we are exposed to? During the past decades, several noise indicators have been developed to predict the negative effects of noise. One main category of noise descriptors expresses noise exposure as an average noise level over a certain time, known as the time equivalent or time average noise level. The 24 h equivalent noise level, $L_{Aeq,24h}$, is an example. This indicator has further developed into a time-weighted noise indicator, L_{den} (day, evening, night), where the evening and night-time periods are given a penalty of 5 dB and 10 dB respectively, to account for a higher sensitivity to noise during these time periods. Furthermore, a specific night-time indicator is also used, L_{night}, which is the time equivalent noise level during the 8-h night-time (e.g., from 23.00 PM to 07.00 AM in most of Europe). L_{den} and L_{night} are the official noise indicators in use by the Environmental Noise Directive, END (Directive 2002/49/EC) that were introduced in 2002. These indicators aim at predicting long-term negative effects of noise, with L_{den} being a general noise indicator and L_{night} used to predict noise-induced sleep disturbance. The other main category of noise indicators seeks to reflect the level of the highest noise event experienced during a specific time-period or for an individual event, the maximum noise level. This $L_{A,max}$, is mainly in use to describe the acute effects on sleep, such as the probability of an arousal or awakening from sleep.

4.2   What is Health?

When we talk about the health effects of noise, it is important to define the term "health". People often relate different meanings to "health", which also varies over time and culture. In 1948, the World Health Organization (WHO) developed the broad definition: *"Health is a state of complete physical, mental and social well-being and not merely the absence of disease or infirmity"*. More recent definitions frame the concept of health in a positive way. Huber et al. (2011), for example, define health as the ability to adapt and self-manage, in the face of social, physical and emotional challenges. In addressing the positive aspects and restorative potential of the environment in this chapter, we utilise this broader definition of health. We consider noise annoyance and sleeping problems that impact well-being and quality of life as falling within the definition of health effects. This is in line with the WHO approach (WHO 2018) that also defined a high degree of noise annoyance and sleep disturbance as a health loss, and thus as adverse effects of noise that should be prevented. We also consider how noise exposure might influence determinants of health, such as physical activity.

4.3 Population Burden—Estimates of Effects

The European Environment Agency recently estimated that in Europe 113 million people are exposed to harmful levels of road traffic noise, 22 million people to harmful levels of railway noise, and 4 million people to harmful levels of aircraft noise (European Environment Agency 2020). These levels of noise exposure are estimated to cause over 12,000 premature deaths, 48,000 new cases of heart disease and 6.5 million to suffer sleep disturbance each year. However, these estimates are likely to underestimate the public health effects of noise, as they only include populations exposed to higher levels of noise (55 dB L_{den}).

4.4 Noise and Health from a Historical Perspective

'The Effects of Noise on Man' by Kryter (1970) is a key book historically within the field. Published in 1970, it focused on the evaluation of the effects of environmental noise on humans, both on the auditory system and on non-auditory effects (so health effects not affecting the function of the hearing organ). The book considered both the definition and measurement of sound, as well as methods for assessing effects. In terms of non-auditory effects, it considered the effects of noise on sleep, pain, and blood circulation.

The 1970s to late 1990s saw studies emerge examining the effects of environmental noise (aircraft, road traffic, railway noise) on children's learning and development (Bronzaft 1981; Bronzaft and McCarthy 1975; Cohen et al. 1973, 1981a; Evans et al. 1995; Hygge et al. 2002), psychiatric hospital admissions (Jenkins et al. 1979; Tarnopolsky et al. 1980), and biological stress responses (Cohen et al. 1981b; Evans et al. 1998). These studies provided important information establishing the effects of noise on health, for informing policy and guidelines. However, learning from studies during this time was sometimes limited by the use of small samples (causing uncertainty in findings), studies being conducted in specific contexts or countries, the use of methods that compared 'high' and 'low' exposure groups which were defined using different exposure thresholds, and the use of different measures for the same outcome. The first WHO community noise guidelines, published in 2000 (WHO 2000), whilst informed by the available evidence and led by experts in the field, took a strong precautionary principal approach to setting guideline values for a range of settings.

Further alignment of methodologies and cross-country studies began in the early 2000s. For example, in Europe, the 2000s saw European Commission funded projects that could examine the effects of environmental noise exposure using the same methodology across countries such as the RANCH (Road traffic and Aircraft Noise and children's Cognition and Health) study of children's health and learning (Stansfeld et al. 2005), and the HYENA (HYpertension and noise Exposure Near Airports) study of hypertension (Jarup et al. 2008). These had larger samples and improved methods. At the same time, research in Germany focusing on cardiovascular effects by Babisch (Babisch 2006; Van Kempen and Babisch 2012), as well as laboratory studies of sleep disturbance, led by Samel and Basner (Basner et al. 2006, 2007, 2008), not only increased knowledge, but also the methodological robustness within the field. Babisch, in particular, undertook some of the first meta-analyses within the field, where the estimates of the effects of noise on heart disease and heart attacks were combined from different studies. This is desirable for policy and guideline development, as effects are estimated across the range of evidence, rather than relying on only one study. The 2010s onwards have also seen increasing use of large-scale longitudinal studies, assessing exposure first and then following individuals over time to see if they develop the outcome of interest. These studies are often based on registry data on disease and, alongside advances in GIS (Geographic Information System), enabled the assessment of long-term exposure to noise as well as estimates of the burden of disease (Sorensen et al. 2011; Cantuaria et al. 2021; Jephcote et al. 2023).

Building the methodological robustness within the field was becoming more apparent by the time the WHO Night Noise Guidelines were published in 2009 (WHO 2009). By 2018 when the WHO Environmental Noise Guidelines for the European Region (WHO 2018) were published, the guidelines were informed by extensive systematic reviews of annoyance, sleep disturbance, cognition, birth effects, mental health, well-being and quality of life, auditory effects, as well as of interventions and applied by the WHO to guidelines development (Guyatt et al. 2008).

4.5 How Do Humans React to Different Types of Noise?

Our nervous system is like a built-in alarm system that is always on the lookout for anything new or potentially important in our surroundings. It has evolved to react more strongly to changes in sensory stimuli than to continuous, unchanging stimulation (Kandel and Jessell 2021). This heightened sensitivity to change has been a huge advantage from an evolutionary standpoint. Think back to our ancestors who roamed the wild. They had to constantly be aware of their surroundings to survive. Any sudden change in their environment could mean the difference between life and death. For example, if they were used to the sounds of a peaceful forest and suddenly heard a new, unfamiliar noise like the growl of a predator, their heightened sensitivity to that change could save their lives. On the other side, our ability to habituate, or get used to continuous and non-threatening stimuli, was advantageous because it helped conserve mental and physical energy. In today's world, the ways our nervous system responds to sensory stimuli continue to play a vital role in how we process and react to sounds in our daily lives. Nevertheless, there are individual differences in how people respond to sound and noise, and beyond the acoustical properties like sound level and intermittency, individual sensitivity, previous experiences as well as the situational context all play important roles.

4.6   Noise as an Environmental Stressor

An environmental stressor is the pressure the environment exerts on a person or population, not the other way around. When someone experiences such pressures, it can lead to feelings of emotional or physical tension, through a biological and psychological response known as stress. Broadly, environmental stressors can come from the outdoor, indoor, food or social environments. A stressor in this context implies any aspect of our surrounding or local environment that is not conducive to good mental or physical health and, in the worst case, may be harmful. Though individual susceptibility also plays an important role (see Chap. 3), many aspects of the environment can be considered stressors. Examples range from poor quality or contaminated air, soil or water and even night-time light pollution, inadequate housing and social support, to human pathogenic diseases. Common to all environmental stressors is that they are largely beyond the control of the individual experiencing them, in addition to being "chronic, negatively valued, nonurgent and physically perceptible" (Campbell 1983).

Just like in the time of our ancestors, sound is an integral part of the communities and the broader environments in which we live, and also one of these known external stressors. These days, there are a multitude of natural and man-made sources of sound – that of running water or bird song, the bustle of a busy city where we might distinguish specific sounds from commerce, construction and transportation, or sounds from nearby neighbours most noticeable in less built-up areas beyond the city hum. When any of these sounds is not wanted by, or is disturbing to an individual, it is considered noise. It is clear that loud sounds can cause a fright, interrupt sleep and at worst damage the hearing organ. Lower levels of sound experienced over more sustained or critical time periods, whether we tune into them or not, however, can also be disturbing to activities and restoration leading to a range of long-term health effects.

4.7   Long-Term Effects of Chronic Noise Exposure

Aside from the effects of loud sounds on the hearing system, there is a broad range of non-auditory effects of noise on health. These non-auditory effects are the focus of this chapter.

Noise by definition is sound that is in some way disturbing. Thus, one of the main effects of long-term exposure to noise is perceived disturbance and annoyance. Other effects most convincingly related to chronic noise exposure include the obvious sleep disturbance, as well as cognitive impairment in children and cardiovascular diseases (Basner and McGuire 2018; Clark and Paunović 2018a; Van Kempen et al. 2018; Smith et al. 2022). Grounded in sufficient scientific evidence, the WHO has identified these as "critical health effects," meaning these are important to consider when assessing issues of noise. A broader range of important long-term health effects of noise are coming to light as research evolves with studies showing that chronic noise exposure is, or may, also be related to metabolic diseases, mental health and neurological disorders and some types of cancer (Gong et al. 2022; Meng et al. 2022; Zare Sakhvidi et al. 2018; Roswall et al. 2023; Sørensen et al. 2021).

4.8 Mechanisms for Long-Term Effects

In the early 2000s, Babisch (2002) introduced the noise reaction model to biologically explain how exposure to noise could impact health. The main premise of the model is that noise is a psychosocial stressor, i.e. an environmental factor interacting with social and cultural factors to influence the mind and behaviour. The cognitive perception of noise is important, most obvious when we think about the typical defensive "fight or flight" response in reaction to alarming noises.

The noise reaction model has two pathways (Fig. 4.1). The direct pathway relates to non-conscious physiological stress due to interactions between the auditory system and the central nervous system. The indirect pathway involves conscious reactions capturing the emotional stress from perceived discomfort (Basner et al. 2014). This means the effects on the human body can happen either as a "reflex" for example when high noise levels are experienced, but also when noise is bothersome or disruptive to activities that require rest, concentration or communication.

Both pathways ultimately lead to a physiological stress response that triggers core bodily systems into action. These systems are responsible for the unconscious control of important body functions including breathing, heartbeat and digestion (autonomic nervous system), and the glands that produce and secrete hormones into the circulatory system (endocrine system). Activation of the sympathetic-adrenal and hypothalamic–pituitary–adrenal axes set off a cascade of events including first the release of stress hormones and dysregulating hormones that control hunger; further, this induces inflammation and oxidative stress. These processes can influence well-known risk factors and precursors to cardiovascular and metabolic disease including increased blood pressure, blood lipid concentration, blood viscosity, inflammatory and blood clotting factors, blood glucose levels, and heart rate variability (Münzel et al. 2021; Recio et al. 2016). Hypertension, ischemic heart disease, stroke, obesity and diabetes are the types of chronic diseases related to these risk factors.

In short, chronic or long-term exposure to noise can disrupt homeostasis—the internal balance or steady state of all body systems to function correctly—jeopardising health in many ways.

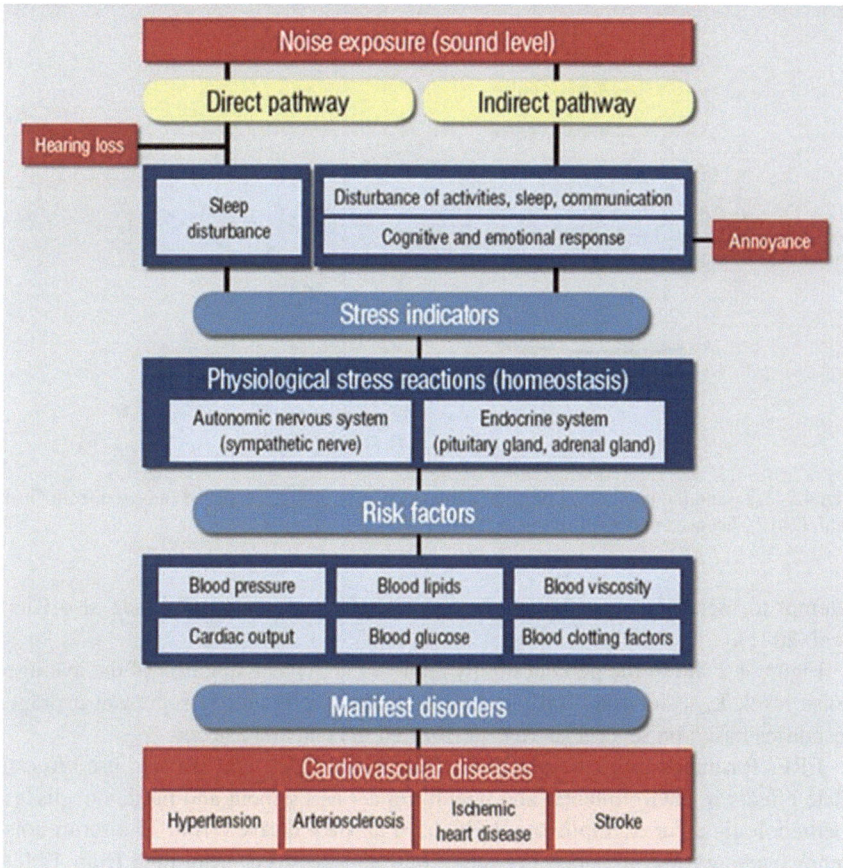

Fig. 4.1 Noise effects reaction scheme. Adapted from Babisch (2002) in Münzel et al. (2014)

4.9 Annoyance Including Non-Acoustic Factors

Noise annoyance encompasses negative reactions to noise such as disturbance, irritation, dissatisfaction and nuisance (Guski 1999), and is one of the most reported community responses in a population exposed to environmental noise. Annoyance is used in policy to measure the quality of life impact of environmental noise exposure on communities, with exposure–response functions (ERFs) plotting the percentage of the population 'highly annoyed' (%HA) against noise exposure using time-averaged metrics for the day or night (e.g., $L_{Aeq, \text{16 h or day}}$, $L_{Aeq, \text{8 h or night}}$). The assessment of annoyance is standardised, following the methodology developed by the International Commission on Biological Effects of Noise (Fields et al. 1997, 2001), as set out in the Technical Standard (ISO, TS15666 2021). The most recent update to the Technical Standard has also set out standardised scoring for %HA, in a further

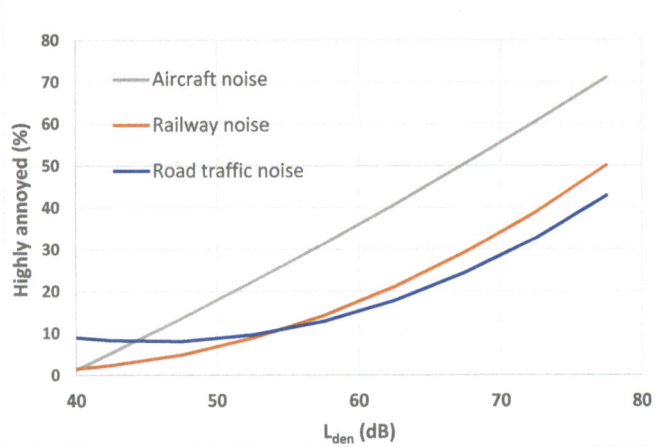

Fig. 4.2 L_{den} and the percentage of highly annoyed (HA) of those exposed (adapted from Guski et al. (2017), licensed under CC-BY 4.0)

attempt to increase comparability between studies internationally (see also Clark et al. 2021).

Figure 4.2 shows the percent highly annoyed (HA) as a function of the averaged noise level, L_{den}, for road, rail, and aircraft noise. The curves represent averaged responses based on several surveys performed in Asia and Europe.

ERFs for annoyance can inform guidance and policy, and are also used to estimate effects in environmental and health impact assessment and burden of disease methodologies. For example, the WHO used an ERF derived from 15 aircraft noise annoyance surveys published between 2000 and 2014 covering data from 17,094 respondents living near very small to international airports, ranging from 34 to 1200 flight movements per day (Guski et al. 2017) to inform their 2018 Environmental Noise Guidelines for the European Region (WHO 2018). The recommended guideline for aircraft noise of $L_{den} = 45$ dB was based on the ERF which estimated that 10% of the population were HA by aircraft noise at that exposure level (see Fig. 4.2). However, there has been debate about the use of annoyance ERFs in this way, given the uncertainty or variability in annoyance levels found for any given sound level across studies. Uncertainty is associated with methodological differences in survey design (sampling, recruitment, population, range of exposure) but also in terms of how noise exposure is estimated, the scoring of annoyance as well as operational differences between airports (e.g. number of runways, night-flights, use of runways and weather), and non-acoustic factors (Clark et al. 2021).

It has long been known that acoustic factors, such as the noise source and sound level, account for only some of the annoyance responses observed. For example, at the same sound level, aircraft noise is most annoying followed by railway and road traffic noise (Guski et al. 2017). Non-acoustic factors, such as attitudes to the noise source (positive or negative), interference with activities, ability to cope, noise sensitivity, expectations, anger, fear, and beliefs about whether noise could be reduced by those responsible influence annoyance responses (WHO 2000). Chapter 3 discusses the role of non-acoustic factors and annoyance in more detail.

Annoyance, per se, may also be a risk factor for poorer health. A recent systematic review and meta-analysis, for example, found strong associations between environmental noise annoyance (as opposed to noise exposure) on depression, anxiety, and general mental health (Gong et al. 2022). Details are further discussed below.

4.10 Sleep

Although sensory impressions are greatly reduced during the sleep period, our brain still processes sounds when we sleep. Noise can also keep us from falling asleep. Numerous studies have demonstrated the sleep-disturbing potential of noise, and disturbance of sleep is acknowledged as one of the more deleterious effects of noise (WHO 2011).

To understand the impact of noise on sleep, it is essential to know a few things about the importance of sleep. Even though the function of sleep is not completely known, it is acknowledged that sleep is fundamental for proper brain functioning, well-being and daily functioning (Drummond and Brown 2001). A typical night's sleep is divided into several cycles, each consisting of distinct stages, including light sleep, deep sleep, and REM (rapid eye movement) sleep. Noise, even at relatively low levels, can fragment these sleep cycles. Frequent awakenings or shifts between sleep stages can prevent individuals from reaching the deep, restorative phases of sleep necessary for physical and mental recovery. In the short run, shortened or fragmented sleep has been associated with daytime sleepiness, and impaired performance including memory, reactivity, and planning (Bonnet 1989). Driver sleepiness causing traffic accidents is an example of how sleepiness can have fatal consequences (Horne and Reyner 1999). Furthermore, sleep disruptions affect the brain's ability to regulate emotions, making individuals more susceptible to mood swings and emotional instability.

Several early studies on noise and sleep were conducted in the laboratory, where responses to playback of recorded environmental noise events at different sound levels were measured by polysomnography (PSG) (Basner et al. 2008, 2011; Griefahn et al. 2008; Smith et al. 2017). PSG is the only measure that indicates whether a person is awake or asleep and provides information about sleep depth, and is reckoned as the gold standard of measuring sleep. Some studies have also been conducted in the home bedrooms of participants where the actual noise situation at home has been measured in parallel with PSG monitoring (Basner et al. 2006; Aasvang et al. 2011; Elmenhorst et al. 2012) providing a more realistic situation to explore the effects of noise on sleep. From such studies measuring acute effects of noise events on sleep, it has been demonstrated that the risk of awakenings, as well as changes from deep to lighter sleep, increases as the maximum noise level for single noise events increases (Basner and McGuire 2018).

To preserve a good night's sleep, it is therefore important to reduce the number of noise events with high noise levels, and several noise guidelines have included recommendations related to the maximum noise level ($L_{A,max}$) during the night time period to prevent noise-induced sleep disturbances (e.g., the previous WHO Community Noise guidelines).

From the early focus on the acute effects of noise events on sleep, more recent research has aimed at strengthening the knowledge of long-term exposure to noise and its impact on health, including the role of sleep. Over time, repeated exposure to noise-related sleep disruptions can accumulate to chronic sleep deprivation. In addition to the fact that sleep is essential for daily cognitive functioning, sleep is vital for the body's ability to regulate blood pressure and maintain a healthy cardiovascular system (Cappuccio and Miller 2017). Furthermore, some evidence suggests that sleep deprivation might lead to dysregulation of a hormone called leptin, which is responsible for regulating appetite and food intake (Reutrakul and Cauter 2018). As a result, noise that causes chronic sleep deprivation can contribute to a host of adverse health consequences, including hypertension and cardiovascular disease as well as obesity and diabetes (Eriksson et al. 2018) (see Sect. 4.8 on Mechanisms). The change in focus from acute to long-term effects of noise on sleep was further motivated by, and enhanced after, the introduction of the Environmental Noise Directive (END) indicators L_{den} and L_{night} (WHO 2018) in which the latter is the annual average noise level during the night-time period. Since then, several studies have aimed at exploring the association between L_{night} and long-term impact on sleep, on both self-reported sleep disturbance such as insomnia symptoms (difficulties falling asleep and frequent awakenings), as well as the use of prescribed sleep medication (Evandt et al. 2017; Roswall et al. 2020). Although not formally standardised such as the annoyance questions, similar questions and answer options have been used to establish exposure–response (ER) functions for night-time noise (L_{night}) and the risk of being highly sleep disturbed (HSD) due to various transport noise sources, such as those developed by Basner and McGuire (2018) as a basis for the WHO noise guidelines and later updated by Smith et al. (2022). In the same way as for high noise annoyance, the ERFs for noise and the probability of being highly sleep disturbed form the basis for health-based recommendations and are used to estimate the public health impact of night-time noise.

Figure 4.3 shows the percentage of persons who are highly sleep disturbed as a function of night-time noise. The curves represent averaged responses to questions on awakenings, difficulty falling asleep, and sleep disturbance due to road, rail, and aircraft noise from several surveys performed in Asia and Europe.

4.11 Cardiometabolic Effects

Noise exposure can stress both the cardiovascular system and metabolism. The cardiovascular or circulatory system includes the heart (cardio) and blood vessels (vascular) and the blood. Its primary role is to carry oxygen, nutrients and hormones

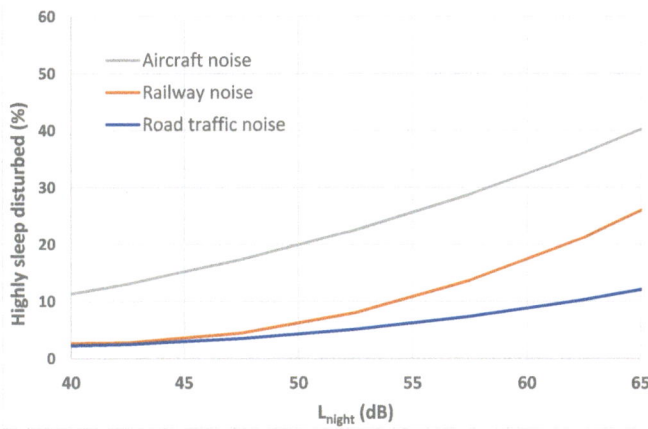

Fig. 4.3 The percent highly sleep disturbed (HSD) as a function of night-time noise level, L_{night} (adapted from Basner and McGuire (2018), licensed under CC-BY 4.0)

around the body. Typical cardiovascular diseases (CVDs) include coronary heart disease (e.g. heart attack) and cerebrovascular disease (e.g., stroke). Metabolism, on the other hand, describes the chemical processes by which nutrients in food are digested and converted to provide energy and other resources to all different organs and their cells. The most common metabolic disease is type 2 diabetes, which influences blood vessel structure and function making it a major contributor to the development of cardiovascular disease. Reflecting these interdependencies, the term cardiometabolic health refers to the (often common) risk factors for developing a cardiovascular or metabolic disease. In essence, diseases of the heart, blood vessels and metabolic system often go hand in hand. The importance of maintaining good cardiometabolic health is well known, and the focus of many interventions and public health messaging. This is because CVDs are the leading cause of death worldwide; diabetes likewise has a large burden, ranked the fourth highest non-communicable disease cause of death worldwide (WHO. The Global Health Observatory. 2023). While the main "modifiable" causes of these diseases are lifestyle choices including diet, alcohol, smoking and exercise (Sattar et al. 2020), environmental exposures including noise also play a role.

A number of studies have shown noise impacts on various cardiometabolic risk factors. Concerning the cardiovascular system, chronic noise exposure is related to stiffening of the arteries, atherosclerosis or narrowing of the arteries, and high blood pressure. Metabolism is influenced by noise through the release of hormones including the stress hormone called cortisol. Cortisol usually helps the body react in a fight or flight response by both inducing the production of glucose and increasing the blood pressure. When stress becomes chronic, sustained elevation of cortisol levels can lead to a chronic increase in blood sugars, insulin resistance, fat deposits in the midsection of the body (central adiposity), and glucose intolerance, all contributing to diabetes. When noise occurs at night, sleep can be fragmented, again leading to stress. Additionally, as we saw above, poor sleep has been shown to impact the balance of hormones responsible for appetite regulation and food intake, possibly further increasing the risk for obesity and type 2 diabetes (Mosavat et al. 2021). Together, these pathways through acute and sustained stress reactions, as well as sleep disturbance, contribute to the development of various cardiovascular and metabolic risk factors.

Early epidemiological research on chronic exposure to transportation noise focused on associations with hypertension and coronary heart disease, with the first cohort study on road traffic noise—in middle-aged men in the UK—and cardiovascular risk factors published in 1988 (Babisch et al. 1988). Twenty years later the influential HYENA study conducted around six major airports in Europe found that those exposed to higher levels of noise had a higher risk of hypertension, both in relation to 24-h road traffic noise and night-time aircraft noise (Jarup et al. 2008). The intervening years saw an increase in the number of studies evaluating if different sources of transportation noise were harmful to cardiovascular health, allowing statistical summarising in so-called meta-analyses to combine the results from diverse populations. Combining data from 24 heterogeneous studies, road traffic noise was shown to increase the risk of hypertension by about 7% per 10 dB(A) increase in day-time noise (Van Kempen and Babisch 2012). Two meta-analyses, with slightly different aims and included studies, calculated a 6 to 8% increased risk of coronary heart disease per 10 dB L_{dn} or L_{den} noise (Babisch 2014; Vienneau et al. 2015). Since then, the WHO Environmental Noise Guidelines (WHO 2018) were devised based on authoritative reviews of both the "critical" and emerging "important" health outcomes related to noise. Most of the evidence was for road traffic noise, and from studies conducted in Europe. Reflecting the evolution of research in the field, the largest body of evidence was on hypertension, gathered mainly from studies using a cross-sectional design (i.e., hypertension measured at one point in time). The highest quality evidence for adverse effects, however, was drawn from a number of longitudinal cohort studies (i.e., following individuals over time to see if they develop the disease) examining ischemic heart disease. A handful of available studies also showed adverse effects for stroke, diabetes and obesity, though this evidence was deemed lower quality at the time due to few studies, in particular a lack of cohort studies.

Since the publication of the WHO guidelines (WHO 2018) the amount and quality of scientific evidence on the cardiometabolic effects of noise has continued to grow including convincing evidence for obesity and diabetes (Münzel et al. 2021). Much of the evidence has been from studies in Europe where detailed exposure prediction from modelling for source-specific noise has been possible for many years. This detail is particularly important in attempting to disentangle the effects of noise from air pollution, both coming from traffic as a common source and both exerting effects on the cardiometabolic system. Important insights beyond those captured in the reviews mentioned above include good indications that the effects of noise on the cardiovascular system are independent of those due to air pollution (Eminson et al. 2023); stronger adverse effects are seen in individuals simultaneously exposed to multiple noise sources (Pyko et al. 2019, 2017; Thacher et al. 2022); night-time noise exposure may be particularly important (Münzel et al. 2020); and even that acute exposure to night-time aircraft noise can trigger cardiovascular deaths (Saucy et al. 2021).

4.12   Children's Learning/Cognition

Research into the effect of environmental noise exposure on children's learning goes as far back as the early 1970s examining the effects of train and subway noise on school performance in New York (Bronzaft 1981; Bronzaft and McCarthy 1975); road traffic noise at home (Cohen et al. 1973); and of aircraft noise at school in the 1980s (Cohen et al. 1981a). Children's learning has been examined by studies that undertake cognitive testing of children's reading comprehension and memory skills, as well as studies that compare children's performance on standardised assessment tests used within education. The RANCH study of over 2000 8–9-year-old children attending schools around London Heathrow, Amsterdam Schiphol, and Madrid Barajas airports found that aircraft noise at school was associated with poorer reading comprehension, as well as with annoyance responses (Stansfeld et al. 2005). A meta-analysis of data from studies around London Heathrow found that a 10 dB increase in aircraft noise at school made children 40% more likely to perform well below or below average on a reading test (Clark et al. 2021). Longitudinal evidence in the field is starting to emerge from cohort studies; a recent study of over 2000 7–10-year-old children found that road traffic noise at school was associated with poorer development of working memory and attention over a one-year period (Foraster et al. 2022).

A study that examined the effect of the closing and opening of the old and new Munich airport found that children who were newly exposed to aircraft noise developed poorer cognition over time, and that cognition improved for those who were no longer exposed to aircraft noise (Hygge et al. 2002). An American study found that performance on standardised test scores improved after insulation works within the school (Sharp et al. 2014), however, more detailed studies that can evaluate the impact of interventions on children's learning are needed, particularly those that might inform guidelines as to which levels of exposure are relevant.

Far less is known about the effects of environmental noise exposure on adult cognition. Recent interest in aging has meant that evidence is starting to emerge for a relationship between environmental noise and dementia and neurodegenerative outcomes (Meng et al. 2022; Clark et al. 2020; Nieuwenhuijsen et al. 2017). A large Danish study found that road traffic noise and railway noise over a 10-year period was associated with the later development of dementia, including Alzheimer's Disease (Cantuaria et al. 2021).

4.13   Mental Health/Neurological Effects

As a stressor, noise is also hypothesised to have a negative effect on mental health, well-being and quality of life (Clark and Paunović 2018b). As described above, short-term exposure to noise results in increased physiological arousal via the endocrine system and autonomic nervous system, leading to an increase in stress hormones like catecholamines (e.g., adrenaline/noradrenaline) and cortisol. If these biological responses are triggered over a long period, they can negatively impact mental health and contribute to the development or retention of depression and anxiety disorders (see below). These biological responses can also be triggered by annoyance and sleep disturbance associated with noise exposure (Gong et al. 2022; Basner and McGuire 2018). Poorer mental health may also make individuals more sensitive to noise.

Mental health has long been a focus in the field of non-auditory effects of noise. As far back as the 1960s, a study by Abey-Wickrama et al. (1969), examining two years of psychiatric hospital admission data showed higher rates of admissions from areas with higher aircraft noise exposure, around Heathrow Airport, compared to areas with lower exposure. Further studies were carried out by Tarnopolsky and colleagues in London in the late 1970s and early 1980s showing that aircraft noise was associated with psychiatric disorders in a community sample, particularly for those with higher education, and importantly, that noise sensitivity was also associated with psychiatric symptoms (Jenkins et al. 1979; Tarnopolsky et al. 1980; Watkins et al. 1981).

However, despite this promising start to the field, relatively few studies of environmental noise effects on mental health have been funded or conducted in recent years, compared to those for other outcomes.

Recent reviews highlight the growing evidence for the effects of environmental noise such as aircraft noise and road traffic noise on mental health, wellbeing and quality of life, cautioning that the evidence and in particular the direction of the associations is uncertain (Clark et al. 2020; Clark and Paunović 2018b; Hegewald et al. 2020). This is an area where further robust high-quality longitudinal studies are needed. There are very few studies of railway noise (Clark et al. 2020); of changes in noise exposure; or evaluations of the impact of interventions on mental health, well-being and quality of life.

The NORAH study, which examined the health insurance data of over 655,000 residents aged over 40 years living near Frankfurt airport, found a relationship between road traffic noise, aircraft noise and railway noise ($L_{Aeq,24 h}$) in 2005 and new cases of clinical depression diagnosed between 2006 and 2010 (Seidler et al. 2017). Similar effects were found for night-time noise exposure. There is also evidence that the use of medication to treat common mental disorders, such as depression and anxiety, increased in areas with higher noise exposure (Floud et al. 2011). A study around Heathrow Airport, examining the impact of a flightpath change, found that there were reductions in prescription spending on nervous and respiratory conditions for regions that experienced a drop in air traffic during the trial (Beghelli 2018).

Studies have suggested that transportation noise (aircraft and road traffic) is associated with hyperactivity symptoms in children (Clark et al. 2021; Schubert et al. 2019).

The changes found are small, and not likely to be clinically significant, increasing the level of symptoms rather than causing illness, per se. An exposure–response function for aircraft noise and hyperactivity symptoms from three studies has been published for use in health impact assessment (HIA) (Clark et al. 2021).

The evidence above has examined how noise as a stressor might lead to the development of poorer mental health, well-being and quality of life. However, there is complexity in the relationship between noise exposure and mental health. Another pathway is that noise annoyance can increase the risk of poorer mental health (Gong et al. 2022). Further, as far back as Tarnopolsky et al. (1980), we have known that mental ill-health increases annoyance (Tarnopolsky et al. 1980; Cerletti et al. 2020). Noise sensitivity has also been shown to increase the effect of noise on mental health, and the effect of poor mental health on response to noise (Cerletti et al. 2020; Stansfeld et al. 2021). Baudin et al. (2021) recently found that both noise annoyance and noise sensitivity were associated with the use of anxiolytic-hypnotic-sedative medication. Just as annoyance is influenced by non-acoustic factors, the effects of noise on mental health and well-being seem influenced by important non-acoustic factors including noise annoyance, noise sensitivity, and pre-existing mental ill-health.

4.14 Cancer

Whether transportation noise is related to cancers was first studied in the Diet, Cancer and Health (DCH) cohort in Denmark. Over 160,000 cancer-free adults were invited to the cohort in the mid-1990s and detailed information including important lifestyle factors and address history were collected. Individuals were followed over time to study if an eventual cancer diagnosis was associated with transportation noise exposure at home (Sørensen et al. 2014). Since then, several other studies—most notably a collaboration to bring together high-quality data from Nordic cohorts well suited to study noise effects—have investigated cancer risk due to noise.

This research has focused on cancer types that have known risk factors that may be perturbed by noise. These include disruption of circadian rhythm, oxidative stress, inflammation, and lifestyle factors including smoking and alcohol consumption. Stress and especially disrupted circadian rhythm (causing inhibited melatonin production) are risk factors for both breast and prostate cancer, while lifestyle factors, obesity and diabetes are more specific risk factors for colorectal cancer.

Most studies have been on breast cancer incidence, generally showing increased risk in relation to transportation noise, while the evidence is less clear and based on fewer studies for prostate and colorectal cancer incidence (Roswall et al. 2023, 2015; Thacher et al. 2023). There is also limited evidence that road traffic noise may be related to cancer mortality in adults (Cole-Hunter et al. 2022; Klompmaker et al. 2021); and recently, one study reported tentative associations between traffic noise and several childhood cancers including Hodgkin lymphoma, non-Hodgkin lymphoma and some CNS tumours (Erdmann et al. 2022). Whether noise influences the progression of disease in those with cancer diagnosis is also of interest; so far, the two studies on patients did not find that noise exposure at home was associated with survival (Roswall et al. 2016, 2017).

4.15 Lifestyle Factors, Socioeconomic Factors and Vulnerable Groups

The increasing use of health impact assessment methodologies and guidelines to protect public health has led to increasing focus on individual differences in effects: namely, might the relationships between noise and health differ for different groups within society? Will general guidelines capture the needs of the most vulnerable?

Those from lower socioeconomic status experience often greater exposure to noise which, alongside increased vulnerability to poorer health, the availability of fewer resources (coping behaviours) and poorer conditions (e.g., poor housing; less access to quiet areas), increases the risk for health-related impacts of noise (European Commission 2016). A recent review by the European Environment Agency (2020) further concluded that "exposure to environmental noise does not affect everyone equally. Socially deprived groups as well as groups with increased susceptibility to noise may suffer more pronounced health-related impacts of noise".

Other groups in the population such as the elderly, shift workers, children, those with pre-existing ill-health, pregnant women, and those who are noise-sensitive might be more vulnerable to the effects of noise on annoyance or other health outcomes (European Environment Agency 2020; Tarnopolsky et al. 1980; Van Kamp and Davies 2013). Reasons for increased vulnerability, as already mentioned, include increased risk for poorer health for these groups. Additionally, these groups tend to spend more time at home and therefore have greater exposure to noise, sleep at times outside of the typical night-time period, and have poorer coping capacities. Some recent studies suggest that specific ethnic groups are exposed to higher levels of environmental noise (Casey et al. 2017; Tonne et al. 2018), but such effects are likely to be context-dependent and need further study, as does the area of vulnerability as a whole. One methodological issue with examining vulnerability is that even large-scale cohort studies often struggle to have the statistical power (i.e., enough people in the sample from the sub-groups of interest) to be able to have certainty in the findings. The context, mentioned above, also means the findings tend to differ across studies and populations. Such issues may need exploring with both quantitative and qualitative approaches.

4.16 Interventions

Despite environmental noise being accepted as a public health issue having significant impacts on the physical health, mental health, and well-being of millions of people, knowledge about effective interventions to protect and promote public health remains scarce (Brown and Van Kamp 2017).

Reducing noise exposure is challenging, as many interventions (e.g., improved home insulation including window glazing, new road surfaces, quieter tyres, electric cars, technological improvements to aircraft) will only individually reduce noise by a few decibels. There is no one, easy fix to reduce transportation noise. It is therefore increasingly important, given urbanisation, that—where interventions are proposed—we understand how they contribute to protecting the public from the non-auditory effects of health.

Establishing and quantifying the effects of interventions on health has long been a quest within the field; as described above for studies of children's learning and health. A recent Swiss study found that a reduction in road speed limit from 50 km/h to 30 km/h was associated with a slight reduction in noise annoyance and sleep disturbance (Brink et al. 2022). Non-acoustic factors are likely to play an important role in the effectiveness of interventions in reducing the health effects of noise.

Evaluating interventions remains a key need within the field. The UK House of Lords Science and Technology Committee inquiry *"The neglected pollutants: the effects of artificial light and noise on human health"* (House of Lords Science and Technology Committee 2023), recently concluded that "research to fill gaps should include the efficacy of interventions to reduce noise pollution on health".

4.17 Positive Soundscapes and Link to Health

Clearly, most research has focused on the negative health effects of noise. More recently the benefits of positive soundscapes (see Chaps. 7, 9) are also of interest in health, planning and policy discussions. High-quality acoustic environments, with pleasant sounds in shared outdoor spaces, are thought to play an important role in well-being and quality of life either intrinsically—because they are pleasing, calming and restorative—or indirectly through buffering us from harmful sound (Van Kamp et al. 2016). Either way, positive soundscapes can help us avoid the annoyance and the stress response evoked by noise.

Daily life, especially if we live in cities, brings us into constant contact with multiple socio-environmental factors. Some are clearly stressors, while others are beneficial and health-promoting including access to services, health care and "islands of tranquillity" in the form of parks and green spaces that foster social cohesion, restoration and physical activity. Growing evidence shows that nature is beneficial for health (Jimenez et al. 2021). National parks and nature reserves, designated and protected by governments and appreciated by visitors, for example, not only offer wondrous sights but also contact with natural sounds (Buxton et al. 2021). Research into the sounds in natural areas, typified by birdsong and fauna, wind, and water, are often positively appraised and perceived as restorative, and capable of reducing stress, fatigue and improving mood (Ratcliffe 2021). In densely populated areas in Asia, where restoration can be more difficult to find amongst the hustle and bustle, there is even a practice known as forest bathing that integrates mediation, walking or simply landscape viewing. Studies have shown that spending time in forests on a regular basis can provide physiological and, in particular, psychological benefits for example by reducing symptoms of anxiety and depression (Siah et al. 2023).

In the European context, green oases within cities also often serve as spaces with lower levels of air pollution. The sounds in these urban green spaces can be natural, or designed to be, pleasant or mask unwanted noise adding to these experiences enhancing relaxation. In a large space, such as a green belt, nature sounds may abound, but even in smaller isolated spaces positive manmade sounds may be added, for example, from a water feature. Urban spaces can thus be purpose designed, through innovative and holistic urban sound planning, to enhance the acoustic and aesthetic qualities of an area to create favourable soundscapes and the visual experience of areas within cities (see Chap. 7). Within a community, this may involve installing green screens (barriers with natural materials, planting trees, installing green facades or roofs) to buffer residential or school areas from major traffic sources, in addition to the interventions mentioned above.

4.18 Summary and Concluding Remarks

This chapter explores the multifaceted impact of noise exposure, extending beyond auditory effects to encompass various non-auditory health implications. Noise, defined as "unwanted and/or harmful sound", is an environmental stressor that disrupts concentration, communication, rest, and sleep, adversely affecting our daily functioning and well-being.

The most studied and well-documented effects of environmental noise are annoyance and sleep disturbance, and noise guidelines are commonly set to protect the general population from being highly annoyed and highly sleep disturbed due to noise. Further understanding of noise and its impact on health, including the physiological mechanism behind, has evolved significantly over the years, especially for transportation noise. With the increasing use of large-scale cohort studies the evidence of the health impact of long-term exposure to traffic noise has profoundly been strengthened. Such studies have established associations between transportation noise and conditions such as ischemic heart disease, stroke, and diabetes as well as adverse effects on children's learning and memory. Research has also indicated associations between prolonged noise exposure and mental disorders, Alzheimer's disease, and certain cancers, but more longitudinal studies of high quality are needed to provide solid knowledge.

Environmental noise is clearly an increasing concern for public health. This chapter underscores the urgency of comprehensive approaches to mitigate the non-auditory health effects of environmental noise exposure. Tailored interventions, informed by diverse research methodologies, are essential. Embracing positive soundscapes and integrating nature into urban environments can foster well-being. As research advances, continued exploration of the intricate interplay between noise and health, including its impact on vulnerable groups, will guide future policies, ensuring healthier and more tranquil living environments for all.

References

Aasvang GM et al. (2011) A field study of effects of road traffic and railway noise on polysomnographic sleep parameters. J Acoust Soc Am 129(6):3716–3726

Abey-Wickrama I et al. (1969) Mental-hospital admissions and aircraft noise. Lancet 2(7633):1275–1277

Babisch W (2002) The noise/stress concept, risk assessment and research needs. Noise Health 4(16):1–11

Babisch W (2006) Transportation noise and cardiovascular risk: updated review and synthesis of epidemiological studies indicate that the evidence has increased. Noise Health 8(30):1–29

Babisch W (2014) Updated exposure-response relationship between road traffic noise and coronary heart diseases: a meta-analysis. Noise Health 16(68):1–9

Babisch W et al. (1988) Traffic noise and cardiovascular risk. The Caerphilly study, first phase. Outdoor noise levels and risk factors. Arch Environ Health, 43(6): pp 407–14

Basner M, McGuire S (2018) WHO Environmental Noise Guidelines for the European Region: a systematic review on environmental noise and effects on sleep. Int J Environ Res Public Health 15(3):519

Basner M, Samel A, Isermann U (2006) Aircraft noise effect on sleep: application of the results of a large polysomnographic field study. J Acoust Soc Am 119(5):2772–2784

Basner M et al. (2007) An ECG-based algorithm for the automatic identification of autonomic activations associated with cortical arousal. Sleep 30(10):1349–1361

Basner M et al (2008) Aircraft noise: Effects on macro- and microstructure of sleep. Sleep Med 9:382–387

Basner M, Müller U, Elmenhorst EM (2011) Single and combined effects of air, road, and rail traffic noise on sleep and recuperation. Sleep 34(1):11–23

Basner M et al. (2014) Auditory and non-auditory effects of noise on health. The Lancet 383(9925):1325–1332

Baudin C et al. (2021) The role of aircraft noise annoyance and noise sensitivity in the association between aircraft noise levels and medication use: results of a pooled-analysis from seven European countries. BMC Public Health 21(1):300

Beghelli S (2018) Health effects of noise and air pollution: empirical investigations. Kings College London

Bonnet MH (1989) Infrequent periodic sleep disruption: effects on sleep, performance and mood. Physiol Behav 45(5):1049–1055

Brink M, Mathieu S, Rüttener S (2022) Lowering urban speed limits to 30 km/h reduces noise annoyance and shifts exposure-response relationships: Evidence from a field study in Zurich. Environ Int 170:107651

Bronzaft AL (1981) The effect of a noise abatement program on reading ability. J Environ Psychol 1:215–222

Bronzaft AL, McCarthy DP (1975) The effect of elevated train noise on reading ability. Environ Behav 7(4):517–527

Brown AL, Van Kamp I (2017) WHO Environmental Noise Guidelines for the European Region: A systematic review of transport noise interventions and their health effects. Int J Environ Res Public Health 14(8):873

Buxton RT et al. (2021) A synthesis of health benefits of natural sounds and their distribution in national parks. Proc Natl Acad Sci U S A, 118(14)

Campbell JM (1983) Ambient Stressors. Environ Behav 15(3):355–380

Cantuaria ML et al. (2021) Residential exposure to transportation noise in Denmark and incidence of dementia: national cohort study. BMJ 374:n1954

Cappuccio FP, Miller MA (2017) Sleep and Cardio-Metabolic Disease. Curr Cardiol Rep 19(11):110

Casey JA et al. (2017) Race/Ethnicity, Socioeconomic Status, Residential Segregation, and Spatial Variation in Noise Exposure in the Contiguous United States. Environ Health Perspect 125(7):077017

Cerletti P et al. (2020) The independent association of source-specific transportation noise exposure, noise annoyance and noise sensitivity with health-related quality of life. Environ Int 143:105960

Clark C, Paunović K (2018a) WHO Environmental Noise Guidelines for the European Region: A systematic review on environmental noise and cognition. Int J Environ Res Public Health 15:285

Clark C, Paunović K (2018b) WHO Environmental Noise Guidelines for the European Region: Systematic review of the evidence on the effects of environmental noise on quality of life, wellbeing and mental health. Int J Environ Res Public Health 15(11):2400

Clark C et al. (2021) Assessing community noise annoyance: A review of two decades of the international technical specification ISO/TS 15666:2003. J Acoust Soc Am 150:3362

Clark C, Crumpler C, Notley H (2020) Evidence for environmental noise effects on health for the United Kingdom policy context: A systematic review of the effects of environmental noise on mental health, wellbeing, quality of life, cancer, dementia, birth, reproductive outcomes, and cognition. Int J Environ Res Public Health, 17(393)

Clark C et al. (2021) A meta-analysis of the association of aircraft noise at school on children's reading comprehension and psychological health for use in Health Impact Assessment. J Environ Psychol, p 101646

Cohen S et al. (1981a) Aircraft noise and children: Longitudinal and cross-sectional evidence on adaptation to noise and the effectiveness of noise abatement. J Pers Soc Psychol 40(2):331–345

Cohen S et al. (1981b) Cardiovascular and behavioral effects of community noise. Am Sci 69(5):528–535

Cohen S, Glass DC, Singer JE (1973) Apartment noise, auditory discrimination, and reading ability in children. J Exp Soc Psychol, 9(407–422)

Cole-Hunter T et al. (2022) Long-term exposure to road traffic noise and all-cause and cause-specific mortality: a Danish Nurse Cohort study. Sci Total Environ 820:153057

Drummond SP, Brown GG (2001) The effects of total sleep deprivation on cerebral responses to cognitive performance. Neuropsychopharmacology 25(5 Suppl):S68-73

Elmenhorst EM et al. (2012) Examining nocturnal railway noise and aircraft noise in the field: sleep, psychomotor performance, and annoyance. Sci Total Environ 424:48–56

Eminson K et al. (2023) Does air pollution confound associations between environmental noise and cardiovascular outcomes?—A systematic review. Environ Res, 2023: p. 116075

Erdmann F et al. (2022) Residential road traffic and railway noise and risk of childhood cancer: A nationwide register-based case-control study in Denmark. Environ Res 212(Pt A):113180

Eriksson C, Pershagen G, Nilsson M (2018) Biological mechanisms related to cardiovascular and metabolic effects by environmental noise. World Health Organanization Regional Office for Europe

European Commission (2016) Links between noise and air pollution and socioeconomic status. In-depth Report 13. Publications Office of the European Union: Luxembourg

European Environment Agency (2020) Environmental Noise in Europe 2020. Publications of the European Union: Luxembourg

Evandt J et al. (2017) A population-based study on nighttime road traffic noise and Insomnia. Sleep, 40(2)

Evans GW, Bullinger M, Hygge S (1998) Chronic noise exposure and physiological-response: A prospective study of children living under environmental stress. Psychol Sci 9(1):75–77

Evans GW, Hygge S, Bullinger M (1995) Chronic noise and psychological stress. Psychol Sci, 6(333–338)

Fields JM et al. (1997) Guidelines for reporting core information from community noise reaction surveys. J Sound Vib 206:685–695

Fields JM et al. (2001) Standardized general-purpose noise reaction questions for community noise surveys: research and a recommendation. J Sound Vib 242:641–679

Fink D (2019) A new definition of noise: noise is unwanted and/or harmful sound. Noise is the new 'secondhand smoke'. 39, 050002

Floud S et al. (2011) Medication use in relation to noise from aircraft and road traffic in six European countries: results of the HYENA study. Occup Environ Med 68(7):518–524

Foraster M et al. (2022) Exposure to road traffic noise and cognitive development in schoolchildren in Barcelona, Spain: A population-based cohort study. PLoS Med 19(6):e1004001

Gong X et al. (2022) Association between noise annoyance and mental health outcomes: A systematic review and meta-analysis. Int J Environ Res Public Health 19(5):2696

Griefahn B et al. (2008) Autonomic arousals related to traffic noise during sleep. Sleep 31(4):569–577

Guski R (1999) Personal and social variables as co-determinants of noise annoyance. Noise Health 1(3):45–56

Guski R, Schreckenberg D, Schuemer R (2017) WHO Environmental Noise Guidelines for the European Region: A Systematic Review on Environmental Noise and Annoyance. Int J Environ Res Public Health 14(12):1539

Guyatt GH et al. (2008) Rating quality of evidence and strength of recommendations GRADE: an emerging consensus on rating quality of evidence and strength of recommendations. BMJ 336:924–926

Hegewald J et al. (2020) Traffic noise and mental health: a systematic review and Meta-Analysis. Int J Environ Res Public Health, 17(17)

Horne J, Reyner L (1999) Vehicle accidents related to sleep: a review. Occup Environ Med 56(5):289–294

House of Lords Science and Technology Committee (2023) The neglected pollutants: the effects of artificial light and noise on human health, in 2nd Report of Session 2022–23. The Authority of the House of Lords

Huber M et al. (2011) How should we define health? BMJ 343:d4163

Hygge S, Evans GW, Bullinger M (2002) A prospective study of some effects of aircraft noise on cognitive performance in schoolchildren. Psychol Sci 13(5):469–474

ISO/TS15666:2021, Acoustics—Assessment of noise annoyance by means of social and socio-acoustic surveys. 2021, International Organization for Standardization: Geneva; Switzerland

Jarup L et al. (2008) Hypertension and exposure to noise near airports: the HYENA study. Environ Health Perspect 116(3):329–333

Jenkins LM et al. (1979) Comparison of three studies of aircraft noise and psychiatric hospital admissions conducted in the same area. Psychol Med 9(4):681–693

Jephcote C et al. (2023) Spatial assessment of the attributable burden of disease due to transportation noise in England. Environ Int 178:107966

Jimenez MP et al. (2021) Associations between Nature Exposure and Health: A Review of the Evidence. Int J Environ Res Public Health, 18(9)

Kandel ER, Schwartz SJ, Jessell TM (2021) Principles of Neural Science. 6th edition

Klompmaker JO et al. (2021) Effects of exposure to surrounding green, air pollution and traffic noise with non-accidental and cause-specific mortality in the Dutch national cohort. Environ Health 20(1):82

Kryter KD (1970) The effects of noise on man. 1st Edition ed

Meng L et al. (2022) Chronic noise exposure and risk of dementia: a systematic review and Dose-Response Meta-Analysis. Front Public Health 10:832881

Mosavat M et al. (2021) The role of sleep curtailment on leptin levels in obesity and diabetes mellitus. Obes Facts 14(2):214–221

Münzel T et al. (2020) Adverse cardiovascular effects of traffic noise with a focus on nighttime noise and the new WHO noise guidelines. Annu Rev Public Health 41:309–328

Münzel T, Gori T, Babisch W, Basner M (2014) Cardiovascular effects of environmental noise exposure. Eur Heart J 35(13):829–36

Münzel T, Sørensen M, Daiber A (2021) Transportation noise pollution and cardiovascular disease. Nat Rev Cardiol

Nieuwenhuijsen MJ, Ristovska G, Dadvand P (2017) WHO Environmental Noise Guidelines for the European Region: A systematic review on environmental noise and adverse birth outcomes. Int J Environ Res Public Health 14(10):1252

Pyko A et al. (2017) Long-Term exposure to transportation noise in relation to development of Obesity—a cohort study. Environ Health Perspect 125(11):117005

Pyko A et al. (2019) Long-term transportation noise exposure and incidence of ischaemic heart disease and stroke: a cohort study. Occup Environ Med 76(4):201–207

Ratcliffe E (2021) Sound and Soundscape in Restorative Natural Environments: A Narrative Literature Review. Front Psychol, 12

Recio A et al. (2016) Road traffic noise effects on cardiovascular, respiratory, and metabolic health: An integrative model of biological mechanisms. Environ Res 146:359–370

Reutrakul S, Van Cauter E (2018) Sleep influences on obesity, insulin resistance, and risk of type 2 diabetes. Metabolism 84:56–66

Roswall N et al. (2015) Residential exposure to road and railway noise and risk of prostate cancer: a prospective cohort study. PLoS ONE 10(8):e0135407

Roswall N et al. (2016) Residential road traffic noise exposure and survival after breast cancer—A cohort study. Environ Res 151:814–820

Roswall N et al. (2017) Residential road traffic noise exposure and colorectal cancer survival—A Danish cohort study. PLoS ONE 12(10):e0187161

Roswall N et al. (2023) Long-term exposure to traffic noise and risk of incident colon cancer: A pooled study of eleven Nordic cohorts. Environ Res 224:115454

Roswall N et al. (2020) Nighttime road traffic noise exposure at the least and most exposed façades and sleep medication prescription redemption—a Danish cohort study. Sleep, 43(8)

Sattar N, Gill JMR, Alazawi W (2020) Improving prevention strategies for cardiometabolic disease. Nat Med 26(3):320–325

Saucy A et al. (2021) Does night-time aircraft noise trigger mortality? A case-crossover study on 24 886 cardiovascular deaths. Eur Heart J 42(8):835–843

Schubert M et al. (2019) Behavioral and emotional disorders and transportation noise among children and adolescents: a systematic review and Meta-Analysis. Int J Environ Res Public Health, 16(18)

Seidler A et al. (2017) Association between aircraft, road and railway traffic noise and depression in a large case-control study based on secondary data. Environ Res 152:263–271

Sharp B et al. (2014) Assessing aircraft noise conditions affecting student learning, A.C.R. Program, Editor. Transp Res Board Natl Acad

Siah CJR et al. (2023) The effects of forest bathing on psychological well-being: A systematic review and meta-analysis. Int J Ment Health Nurs 32(4):1038–1054

Smith MG et al. (2017) Physiological effects of railway vibration and noise on sleep. J Acoust Soc Am 141(5):3262

Smith MG, Cordoza M, Basner M (2022) Environmental noise and effects on sleep: an update to the WHO systematic review and Meta-Analysis. Environ Health Perspect 130(7):76001

Sorensen M et al. (2011) Road traffic noise and stroke: a prospective cohort study. Eur Heart J 32(6):737–744

Sørensen M et al. (2014) Exposure to road traffic and railway noise and postmenopausal breast cancer: A cohort study. Int J Cancer 134(11):2691–2698

Sørensen M et al. (2021) Road and railway noise and risk for breast cancer: A nationwide study covering Denmark. Environ Res 195:110739

Stansfeld SA et al. (2005) Aircraft and road traffic noise and children's cognition and health: a cross-national study. The Lancet 365(9475):1942–1949

Stansfeld S et al. (2021) Road traffic noise, noise sensitivity, noise annoyance, psychological and physical health and mortality. Environ Health 20:32

Tarnopolsky A, Watkins G, Hand DJ (1980) Aircraft noise and mental health: I. Prevalence of Individual Symptoms. Psychol Med 10(4):683–698

Thacher JD et al. (2022) Exposure to transportation noise and risk for cardiovascular disease in a nationwide cohort study from Denmark. Environ Res 211:113106

Thacher JD et al. (2023) Exposure to long-term source-specific transportation noise and incident breast cancer: A pooled study of eight Nordic cohorts. Environ Int 178:108108

Tonne C et al. (2018) Socioeconomic and ethnic inequalities in exposure to air and noise pollution in London. Environ Int 115:170–179

Van Kamp I, Davies H (2013) Noise and health in vulnerable groups: a review. Noise Health 15(64):153–159

Van Kamp I, et al. (2016) Chapter 3: Soundscapes, Human Restoration, and Quality of Life, in Soundscape and the built environment, Brown AL, Gjestland T, Dubois D, Editors. p 1–16

Van Kempen E, Babisch W (2012) The quantitative relationship between road traffic noise and hypertension: a meta-analysis. J Hypertens 30(6):1075–1086

Van Kempen E et al. (2018) WHO Environmental Noise Guidelines for the European Region: a systematic review on environmental noise and cardiovascular and metabolic effects: a summary. Int J Environ Res Public Health 15(2):379

Vienneau D et al. (2015) The relationship between transportation noise exposure and ischemic heart disease: A meta-analysis. Environ Res 138:372–380

Watkins G, Tarnopolsky A, Jenkins LM (1981) Aircraft noise and mental health: II. Use of medicines and health care services. Psychol Med, 11(1): pp 155–68

WHO (2000) Guidelines for Community Noise. World Health Organization Europe: Geneva, Switzerland

WHO (2009) Night Noise Guidelines for Europe. World Health Organization Europe: Cophenhagen, Denmark.

WHO (2011) Burden of Disease from Environmental Noise. World Health Organization, Europe: Copenhagen; Denmark.

WHO (2018) Environmental Noise Guidelines for the European Region. World Health Organisation, Regional Office for Europe: Copenhagen, Denmark

WHO. The Global Health Observatory. 2023; Available from: https://www.who.int/data/gho/data/themes/topics/topic-details/GHO/ncd-mortality

Zare Sakhvidi MJ et al. (2018) Association between noise exposure and diabetes: A systematic review and meta-analysis. Environ Res 166:647–657

Chapter 5
Determining the Population Health Impact of Environmental Noise

Mark Brink and Juanita Haagsma

Noise is the most impertinent of all forms of interruption. It is not only an interruption, but also a disruption of thought. Arthur Schopenhauer.

5.1 Introduction

Long-term exposure to environmental noise, such as road traffic or railway noise, can result in severe health consequences. In the EU in terms of the burden of disease (BoD), environmental noise is estimated to be the second most important environmental risk factor after air pollution. At least 20% of the EU population lives in areas where road traffic noise levels are considered to be harmful to health (European Environment Agency (EEA) 2020). As described in several parts of this book (in particular, in Chaps. 3, 4 and 7), a range of unwanted health effects can be attributed to environmental noise. Examples of such effects are the number or percentage of persons experiencing annoyance due to noise (e.g. suffering from high annoyance (Guski et al. 2017) or from noise-induced sleep disturbance (Brink et al. 2019)), reduced functioning in usual activities (e.g. reduction in reading performance among children exposed to railway noise (Klatte et al. 2013)), health care consumption (e.g. number of hospital admissions due to cardiovascular disease (Correia et al. 2013), cardiovascular and metabolic morbidity (Munzel et al. 2021; Kempen et al. 2018), and of course, mortality (Vienneau et al. 2022). These endpoints can provide a first indication of the extent of the impact of noise exposure on public health. However,

M. Brink (✉)
Federal Office for the Environment, Bern, Switzerland
e-mail: Mark.Brink@bafu.admin.ch

J. Haagsma
Department of Public Health, Erasmus MC University Medical Center, Rotterdam, Netherlands
e-mail: j.haagsma@erasmusmc.nl

© The Author(s) 2025   75
I. van Kamp and F. Woudenberg (eds.), *A Sound Approach to Noise and Health*,
Springer-AAS Acoustics Series, https://doi.org/10.1007/978-981-97-6121-0_5

because noise exposure may result in diverse adverse health effects in the population under study, an integrated approach covering all relevant endpoints seems most appropriate to quantify the overall health burden in entire populations.

An effective noise abatement policy should allocate resources in a way that maximizes benefits while avoiding undue interference with other societal functions and human activities. Methods referred to as "health impact assessment" can be used to provide the relevant numbers in such contexts. A health impact assessment is the systematic evaluation of potential adverse health effects resulting from exposure to a particular environmental factor – in our case, environmental noise. The main purpose of such an assessment is to estimate the health impact of exposure to noise or changes in noise in different socioeconomic, environmental and policy settings. Health impact assessments are often also a necessary intermediate step towards the economic quantification of noise impacts (see Chap. 6). The corresponding information is essential when it comes to the allocation of funds, monitoring population health, development of prevention measures or health interventions and the evaluation of the effect of their implementation.

In general, there are five steps in health impact assessment due to environmental noise:

1. Definition of an exposure scenario for which health effects should be calculated. This includes decisions about the environmental noise sources to be included, the year, the geographical region and the choice of a so-called reference scenario (also known as "counterfactual"). The reference scenario for health impact assessment is usually a situation 'without' environmental noise or with the theoretical minimum exposure level, i.e. only low levels of noise.
2. Estimation of the distribution of environmental noise exposure (of the relevant source(s)) for the target population.
3. Decision on a specific set of health endpoints to be included and evaluation of the association between exposure and response for the selected endpoints. This is often achieved by referring to systematic literature reviews or meta-analyses that provide the necessary exposure – response relationships.
4. Collection of baseline health data for the selected health endpoints, which are needed as ancillary inputs, e.g. general myocardial infarction mortality risk in the population (for the outcomes high annoyance and/or high sleep disturbance, this is not necessary, because the corresponding figures are always expressed as absolute risks per exposure category, hence directly applicable).
5. Quantification of the impact in terms of the number of attributable cases/deaths, number of years of life lost, years lost to disability, or so-called Disability Adjusted Life Years (DALYs). DALYs will be discussed in more detail below.

General principles for the conduct of health impact assessment and quantifying the burden of disease, of relevance for environmental factors, can be found in (Knol et al. 2009; Joffe and Mindell 2005; Van Kamp et al. 2018).

5.2 Health Impact Indicators, Indexes, and DALYs

5.2.1 Overview

An array of health indicators can be used to quantify the health impacts of noise exposure. They range from the number of people exposed to noise levels above a certain threshold, e.g. a national noise exposure limit value, or a guideline value by the WHO (2018), to more complex indicators that reflect the number of people suffering from specific health outcomes or that combine multiple health effects into a single figure, like the Disability Adjusted Life Years (DALY). The suitability of one indicator over another depends on the policy questions to be answered, the audience to which the results will be communicated, and also the availability of baseline health data, exposure data, manpower and expertise (Kamp et al. 2018). Van Kamp et al.'s report (2018, p. 15) contains a comprehensive table of the most relevant population-based indicators. Not all indicators are easily understandable by the general public, and some indicators at first glance seem to be better suited to make hesitant politicians realize that noise is really dangerous for public health (e.g. mortality figures), than others (e.g. number of annoyed persons). Typical indicators of population health are cause-specific mortality rates, numbers of new (incident) and existing (prevalent) cases of a disease, or self-reported percentage of highly annoyed persons (%HA). There are also other probably less known integral noise impact assessment instruments that are used in specific policy contexts, e.g. so-called aircraft noise indexes, such as the Zurich Aircraft Noise Index (ZFI) and Frankfurt Aircraft Noise Index (FFI). These instruments are, similar to the DALY, a sort of single figure marker that represents the total accumulated undesired effects of aircraft noise generated by a particular airport in a given area. They have been implemented to make the long-term effects of operational decisions (e.g. routes, fleet, number of movements, etc.) visible and comparable to each other on the population level (Brink et al. 2010).

When faced with so many different types of outcome measures, it becomes challenging to determine where resources (e.g. health care or noise abatement resources) should be most efficiently directed. In order to fully and integrally describe the public health effects of environmental noise, a metric would be needed that combines the fatal and non-fatal health effects into one number. A unit for such a number that is often used for this purpose is time or time duration. Examples of widely used summary measures of population health are Healthy Adjusted Life Years (HALYs), Quality Adjusted Life Years (QALYs) and Disability Adjusted Life Years (DALYs), or the Cumulative Health-based Environmental Risk Indicator (CHERIO). Summary measures of population health like the DALY are an important tool for priority setting in public health decisions due to their feature of aggregation of both fatal and non-fatal health effects and translating them into the same unit (time). This allows for the determination of the full impact of a certain disease, injury or risk factor on population health and comparison with other diseases, injuries and risk factors, comparison across time and comparison across regions. These metrics can also be used to quantify noise effects in populations. In the following, we will shed some light on the most common of these metrics, the DALY.

5.2.2 Calculation of DALYs

The DALY metric is a health gap measure that integrates the healthy time (years) lost due to mortality and morbidity into a single figure (Field and Gold 1998a, 1998b). This allows comparison of the burden of disease across a range of illnesses, injuries, risk factors, interventions or populations (Murray et al. 2002). Studies show that the use of DALYs has certain advantages over conventional environmental impact assessment for quantification and comparison of the risks resulting from environmental pollution (Gao et al. 2015). The DALY was first introduced by the World Bank in 1993 and since its application in the 1996 Global Burden of Disease (GBD) study (Murray and Lopez 1996; Worldbank 1993), the DALY has been applied widely in priority settings in health care and prevention. The GBD study is an ongoing investigation that determines mortality, incidence, prevalence, life expectancy, healthy life expectancy (HALE) and DALYs for over 300 diseases and injuries and more than 80 risk factors, including environmental risk factors, e.g. in- and outdoor air and particulate matter pollution, spanning 204 countries and territories (Murray et al. 2012; GBD 2019). However, the GBD study does not include environmental noise as a risk factor yet. This means that currently there is no information on a truly global level of DALYs attributable to environmental noise.

The DALY summarizes health loss at the population level into a single index by adding a) years of life lost (YLL), and b) years lived with disability (YLD) (Murray and Acharya 1997). Thus, DALY = YLL + YLD. While the DALY concept has broad appeal, several methodological challenges in deriving correct estimates for specific populations, or even on a global scale, are evident and are addressed in Paragraph 4.

YLL represents the time lost through premature mortality and is calculated with the following formula:

$$YLL = \sum d_1 * e_1$$

where d are the number of fatal cases, due to health outcome 1 and e is the expected individual life span at the age of death.

YLD represents the healthy time lost while living with a disability and is calculated with the following formula:

$$YLD = \sum n_1 * t_1 * DW_1$$

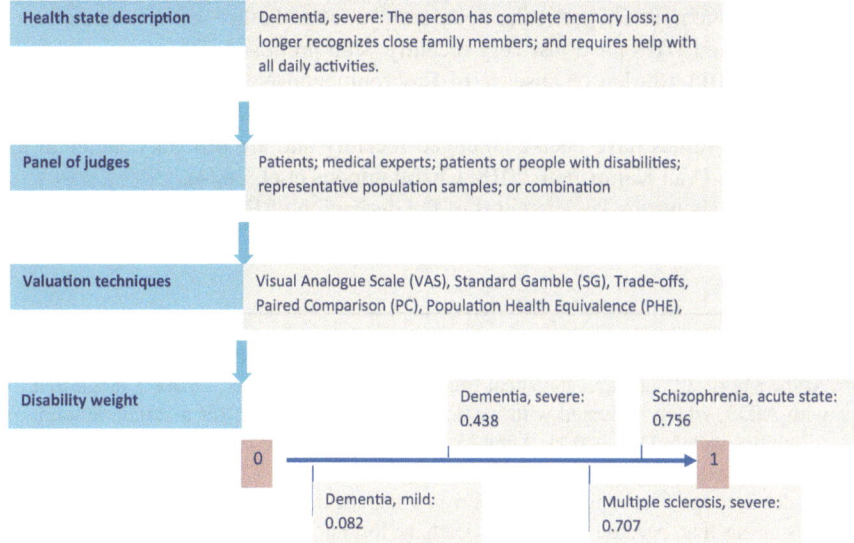

Fig. 5.1 Key elements of the derivation of disability weights (DWs)

where n are the number of cases with health outcome l, t is the duration of the health outcome l and DW is the disability weight of health outcome l. The DW is a value assigned to living with a disease or injury and it is anchored between 0 (equivalent to "perfect health") and 1 (equivalent to "death") and represents the impact of a disease or injury health state on a scale, as a percentage reduction from full health. For instance, if the DW for long-term consequences of stroke was 0.50, then the conditions of two people living with long-term consequences of stroke in the year of interest would be equivalent to the loss of one year due to premature mortality.

Where do DWs come from? DWs are based on the health state evaluations of a group of people (Haagsma et al. 2014). These people can consist of health experts or members of the general population. Conceptually, this means that DWs do not reflect any kind of self-reported experiences with disability or illness by people affected by a certain condition, but rather describe a kind of shared consensus (either by experts or lay people) on the severity of a certain illness, disease or disability relative to the severity of other illnesses, diseases or disabilities. The issue of who should value health states is still contentiously debated in the literature. The incorporation of the views of the general public, i.e. laypersons, has been recommended for deriving DWs, because the burden of disease studies is primarily used as a tool for guiding decision-making on resource allocation at the population level (Field and Gold 1998a, 1998b). In the past, several DW measurement studies have been performed (Charalampous et al. 2022). It is not difficult to see that DWs carry great weight in the calculation of DALYs.

Disability weights for a range of health outcomes have been available since 1996 (Murray and Lopez 1996). However, for some of the outcomes caused by or contributed to noise, in particular noise annoyance and other self-reported effects, robustly derived DWs have just very recently been proposed (Charalampous et al. 2024). The 2011 Burden of Disease of Environmental Noise report by the WHO (2011) employed a DW of 0.02 for high annoyance and 0.07 for high sleep distur-bance. These values have been challenged recently and are probably rather in the region of 0.01 (Van Kamp et al. 2018; Charalampous et al. 2024).

Figure 5.1 shows the key elements of the derivation of DWs.

Box

Example of a DALY calculation in one individual person's life:

At the age of 30, having a statistical life expectancy of 77 years, a man is diagnosed with AIDS, which is treated with medication but disables him for a certain amount (disability weight DW = 0.1). After 25 years, at 55, he starts to suffer from another more severe disease (DW = 0.3) for 10 years (a total DW of 0.4 during 10 years) before his death at age 65. At age 65, this person's statistical life expectancy was 80 years. This means that, because of his early death, he lost another 15 years of healthy life (a DW of 1.0 during 15 years). In total, in this case, 21.5 disability-adjusted life years were lost (Fig. 5.2).

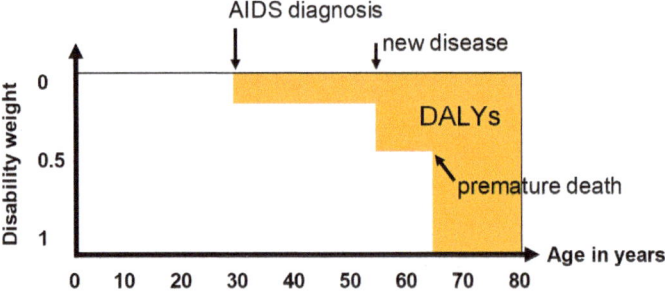

Fig. 5.2 Calculation of disability adjusted life years (DALY) for a hypothetic individual

There are of course, less drastic examples, in particular those related to the health effects of noise: say, a young student moves from the countryside to the city to study. She lives in a flat shared with two colleagues along a busy road for 4 years before she moves to a quieter place on-campus shortly before she starts with her PhD project. At the beginning of her studies, she felt that she would easily get used to the road traffic noise but this was not the case and she remained highly annoyed by the noise during the first four years at the university. Assuming a DW of just 0.01 for this health status (namely, to be highly annoyed by noise), the equivalent loss expressed in DALYs is 0.04, or – roughly – two weeks of healthy life lost for her within just four years. We see that, put into perspective, despite small DWs, noise can have a considerable effect on our quality of life that should not be neglected.

In order to determine DALYs attributable to a risk factor, such as environmental noise, the first step is to determine for which health outcomes there is both a plausible biological pathway as well as sufficient scientific evidence that exposure to the risk factor is associated with the health outcome. Secondly, to assess the health effect of a given exposure, a pertinent exposure-effect relationship is necessary. The scientific evidence on the strength of the association between risk and outcome is compared to pre-defined criteria (e.g. by the GRADE approach (Cochrane Collaboration 2023)) to include only viable risk-health outcome pairs in the calculations.

For the selected risk factor-health outcome pairs (e.g. noise exposure expressed as Lden and ischaemic heart disease (IHD)), exposure – response relationships have to be determined. A frequently used source for this are systematic literature reviews of cross-sectional or longitudinal studies that collect information on risk factor exposure levels and occurrence (e.g. incidence or prevalence) of the health outcome. Based on the information from such studies, an exposure – response relationship can be determined. The best examples of such systematic reviews have been carried out in connection with the development of the WHO Environmental Noise Guidelines and the discussion and update of its empirical foundations in the aftermath of their publication. Noteworthy are, e.g., the work on noise-induced sleep disturbances by Smith et al. (2022), the systematic review and meta-analysis of 21 studies about road and railway traffic noise exposure and annoyance (Fenech and Clark 2018), or the elucidation of the relationship between environmental noise and cardiovascular and metabolic effects (Van Kempen et al. 2018).

The second step is to calculate the attribution of DALYs to the risk factor under study. A frequently used method for this is comparative risk assessment (CRA). With CRA, the level of exposure in a population is estimated based on data sources on the level of exposure and with as much detail on region and other variables, such as age and gender, as possible. Data on population exposure to environmental noise are usually provided by local or national authorities. For example, in the EU, the member states are obliged by the Environmental Noise Directive (END) to report the number of people exposed to various levels of road traffic, rail, aircraft and industrial noise. Lastly, also the level of exposure associated with a minimum risk must be determined (the so-called "counterfactual (scenario)", see also Sect. 4.3 in this chapter). This level of exposure is referred to as the theoretical minimum risk exposure level.

The third step is to multiply population-attributable fractions (PAFs) of the risk factor of interest by the DALYs of the outcome of interest (e.g. cardiovascular disease) for each region and population subgroup. This multiplication results in the number of DALYs of the outcome of interest that are attributable to the risk factor of interest. For example, the number of DALYs related to cardiovascular disease that are attributable to environmental noise. Next, the DALYs for all selected health outcomes for noise exposure are added together to finally determine the burden of disease of environmental noise (Fig. 5.3).

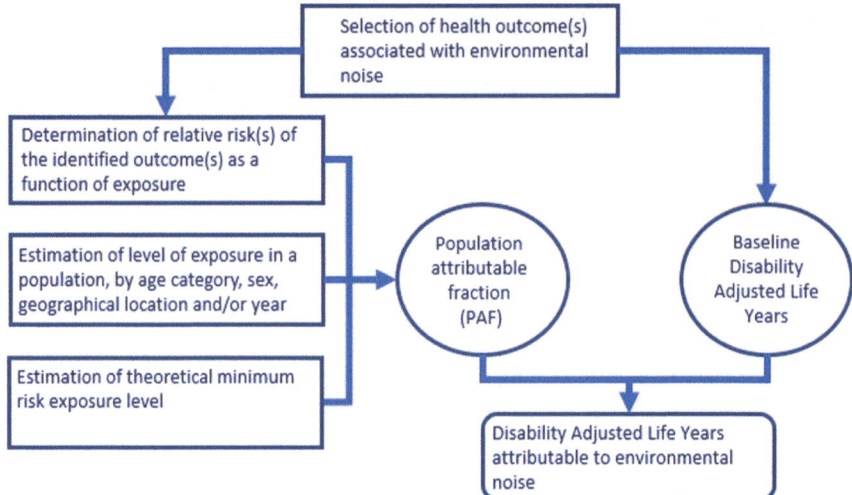

Fig. 5.3 Flow diagram of the steps involved in determining DALYs attributable to environmental noise

5.3 Overview of Health Impact Assessments of Noise

5.3.1 Burden of Disease from Environmental Noise, WHO 2011

In its 2011 report "Burden of disease from environmental noise: Quantification of healthy life years lost in Europe", the WHO for the first time used DALYs to quantify noise effects on health (WHO 2011). This report has become a true citation classic nowadays and virtually no introductory chapter of any review or overview paper on noise effects seems to do without it. However, it is nowadays probably slightly outdated, mainly for reasons that will be discussed further below. The WHO 2011 report aimed at providing technical support to noise mitigation authorities in the quantitative risk assessment of environmental noise, using evidence and data available in Europe. With – maybe aside from the DWs adopted – conservative assumptions applied to the calculation methods, the WHO estimated that DALYs lost from environmental noise are 22 000 years for tinnitus, 61 000 years for ischaemic heart disease, 45 000 years for cognitive impairment of children, 903 000 years for sleep disturbance, and 654 000 years for annoyance in the European Union member states and other western European countries (Fig. 5.4). Sleep disturbance and annoyance, mostly related to road traffic noise, comprised the main burden of environmental noise. While these are certainly not severe health outcomes, the large number of people affected, illustrated by the use of DALYs, demonstrates the relevance of these outcomes for public health. If all the effects above are considered together, the total burden of health effects from environmental noise would be greater than one million

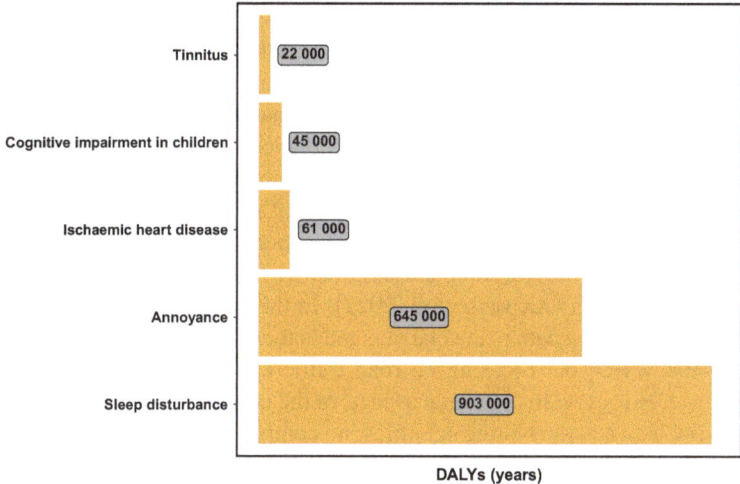

Fig. 5.4 DALYs lost due to environmental noise in Western Europe, as calculated by the WHO (2011)

years in Western Europe. This places noise pollution as the second most burdensome environmental hazard after air pollution.

5.3.2 Regional and National Noise Impact Assessments

Stirred by the release of the WHO report and later the publication of detailed technical guidance for the assessment of the burden of disease from environmental noise (WHO 2012), other countries and/or regions have carried out their own environmental noise impact assessment calculations, that partly reported DALYs, and continue to do so. The ICBEN 2023 review on policy and economics (Fenech and Janssen 2023) identified the most recent environmental noise health impact assessments in the years 2021 and 2022: they were carried out in a study covering more than 700 European Cities (Khomenko et al. 2022), in the Bundesland of Hessen, Germany (Hegewald et al. 2021), in Estonia (Veber et al. 2022), and even in Teheran (Shamsipour et al. 2022). The vast analysis by Khomenko et al. focusing on 724 European cities and 25 capital cities estimated that approximately 11 million persons are highly annoyed by road traffic noise and that 3608 deaths from Ischaemic Heart Disease (95% CI: 843–6266) could be prevented annually if noise levels would comply with the WHO recommendations (Khomenko et al. 2022). We currently do not know about any large-scale noise-related health impact assessments covering the USA or parts of it.

The Swiss Federal Office for the Environment issued DALY calculations for the whole of Switzerland, which were published a few years after the release of the WHO Burden of Disease Report (Ecoplan 2019). Ischaemic heart disease, sleep disturbance and annoyance were the outcomes considered. The calculations were based on new exposure data as well as the most recent estimates of exposure – response functions published by the WHO (2018). Overall, in Switzerland (population 9 million), some 69 300 DALYs are lost due to transportation noise related health outcomes per annum. Most of the DALYs were related to sleep disturbance (44%) and annoyance (43%). Road traffic noise was responsible for the majority of DALYs lost. Rail and aircraft noise accounted for just 13% and 10% of the DALYs, respectively.

A large group of researchers from the Nordic countries recently reported on their DALY assessment for the Nordic countries (Norway, Denmark, Sweden, and Finland) and their capital cities (Aasvang et al. 2022). In this study, the authors followed a similar methodological path to the one pursued in the Swiss calculations and initially paved by the WHO, but focusing on road traffic and railway noise and omitting aircraft noise. However, since several aspects of the noise exposure modelling varied considerably between the Nordic countries, no comparable estimates could be made for the entire countries but just for their capitals. Comparable estimates (for the capitals) ranged from 330 to 485 DALYs per 100 000 inhabitants for road traffic noise, and from 40 to 140 DALYs per 100 000 for railway noise. High annoyance and high sleep disturbance accounted for the largest part of DALYs, in line with the assessments by the WHO.

Whereas DALY calculations are obviously popular, non-summary measures to describe the noise impact on health were also often used in recent studies. The European Environment Agency carried out their own reckonings for 33 partnering European countries (also including Non-EU member states), which they published in 2020, with the following results: Long-term exposure to environmental noise is estimated to cause 12 000 premature deaths and contribute to 48 000 new cases of ischaemic heart disease per year. They estimated that 22 million people suffer chronic high annoyance and 6.5 million people chronic high sleep disturbance due to noise (European Environment Agency (EEA) 2020).

5.3.3 Putting Noise Effects on Health into Perspective

One question that is of course of interest to both epidemiologists and noise policy-makers concerns the health damage caused by noise in comparison to the overall health burden or in comparison to other environmental pollutants. Noise pollution certainly deserves the label "deadly". For example, in Switzerland, with a population of about 9 million, it is estimated that around 450 cardiovascular deaths occur annually due to road traffic noise (Röösli et al. 2019). While this is about twice as high as the average death toll caused by road traffic accidents *on and of that same road network!* – the figure is of course not as dramatic if we consider the total number of cardiovascular deaths in the country, namely about 20,000 (Source: Swiss Federal Statistical Office). Nonetheless, if you do the math, more than 2% of all cardiovascular deaths are attributable to road traffic noise, which may come as a surprise to some.

In a comprehensive analysis from 2016, Prüss-Üstün et al. for the WHO estimated that almost a quarter of global deaths and/or DALYs are attributable to harmful influence ("noxae") from the environment (Fig. 5.5) (Prüss-Üstün et al. 2016). Noxae were defined as "all the physical, chemical and biological factors external to a person, and all related behaviors, but excluding those natural environments that cannot reasonably be modified". In that analysis, stroke and ischaemic heart disease have the highest share of environmental causation (42% and 35% respectively). The report mentions noise, but lacks the reporting of exact figures (certainly because noise exposure data are impossible to get hold of on a global scale), so that we have to consider older (and spatially "narrower") literature to get an idea of the order of magnitude of health damage that noise causes, compared to other environmental exposures (such as air and water pollution, or indoor radon, to name just a few). An earlier study expressing disease burden as DALYs was carried out in the Netherlands and puts noise exposure right after air pollution, with 24 and 59% of the total environmental disease burden (De Hollander et al. 1999). Hänninen and his team for the EEA reckoned that transportation noise in 6 European countries (Belgium, Finland, France, Germany, Italy, and the Netherlands) accounts for about 8% of the burden of disease from a selection of environmental noxae (Hanninen et al. 2014), with PM2.5 (fine particulate matter) being the pollutant with the highest share (68%) (Fig. 5.6).

The relative contribution of environmental factors in these European countries is certainly different from middle- and lower-income countries, and is likely to change over the years, e.g. with the fade-out of carbon-fueled individual transport in cities.

Fig. 5.5 DALYs attributable to environmental and non-environmental risk factors on the global scale, assessed by the WHO for the year 2022. Sources: (Hanninen et al. 2014; Prüss-Üstün et al. 2016)

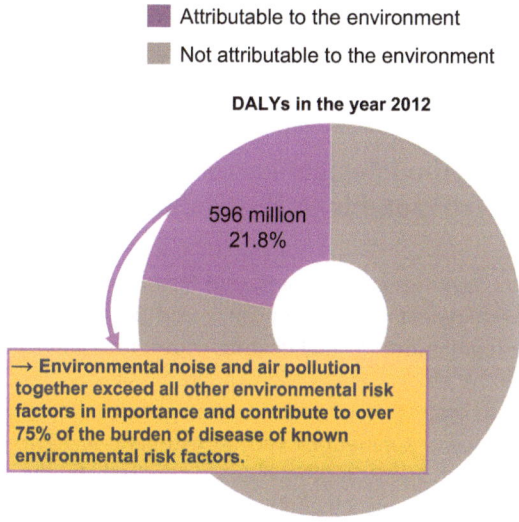

Attributable to the environment

Not attributable to the environment

DALYs in the year 2012

596 million
21.8%

→ Environmental noise and air pollution together exceed all other environmental risk factors in importance and contribute to over 75% of the burden of disease of known environmental risk factors.

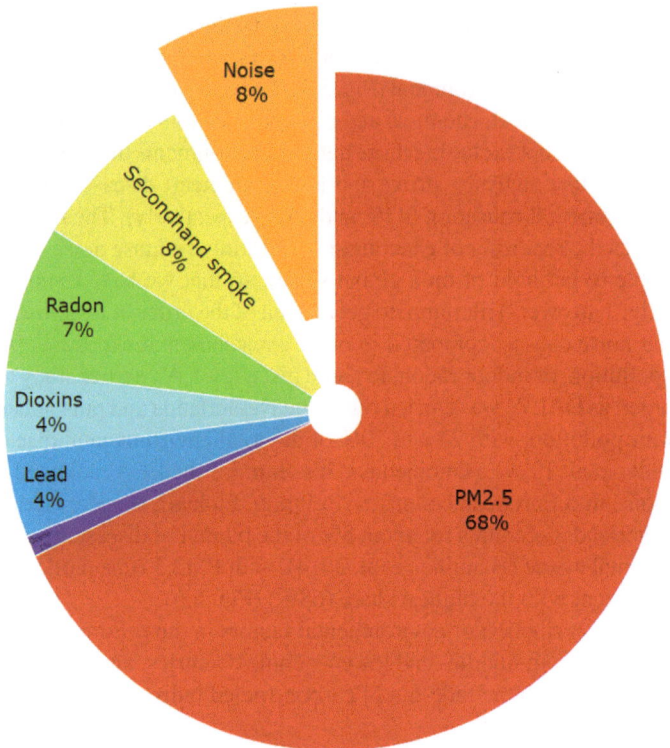

Fig. 5.6 Percentage contribution of nine environmental risk factors to the estimated environmental burden of disease in six countries. Numbers reported by Hänninen et al. (2014). (Not shown: Formaldehyde and Benzene which each contribute less than 1%)

5.4 Methodological Issues of Estimating Total Health Risks Attributable to Environmental Noise

When carrying out environmental health impact assessments, there are a range of methodological issues to consider and decisions to be taken, and the handling of uncertainties and optimal dealing with shortcomings in the data availability demand respective attentiveness. Most of these issues are not specific to noise exposure, but relate to health impact assessment in general. Recurrent sources of error include the inaccuracy of population exposure estimates and limited information on the shape of exposure – response relationships, particularly for the lower exposures. The following sections focus on such challenges and potential solutions.

5.4.1 Choosing Suitable Outcomes/Endpoints

The publication of noise effects research papers has seen a steep increase after the years based on which the WHO's guideline values – roughly 2000 to 2015 – have been published. This led to the situation that the WHO guidelines from 2018 (and their exposure – response relationships) were already outdated in certain aspects at the time of their publication, especially with regard to outcomes that had not yet been investigated or had only been investigated to a limited extent. Some of these 'novel' effects have extensively been discussed on the occasion of the 14th ICBEN Congress on noise as a public health problem in 2023 (Sorensen 2023). In essence, today, more up-to-date systematic reviews and meta-analyses can inform the choice of health effects for noise impact assessments than in the past. Röösli et al. recently proposed to consider diabetes, overweight, and mental health problems including depression, to be included in noise health risk assessments (Röösli et al. 2022). Also, so far only rather narrow diagnostic criteria have been used in noise effect studies, in particular compared with air pollution studies, where, e.g. all-cause mortality is a common outcome for BoD calculations. Studies on the effect of noise exposure on all natural causes of mortality do exist but are still relatively rare (Thacher et al. 2022).

Another important question which was debated in the past is whether "annoyance" should be included as an outcome in health impact assessments. Several arguments can be found in the literature as to why noise annoyance should not be included, with the most important being that annoyance does not have a diagnostic code in the International Statistical Classification of Diseases and Related Health Problems (ICD). However, there are also good arguments in favor of including annoyance, since annoyance can well be seen as an endpoint in its own right irrespective of functional (medical) aspects, which is – by the way – also true for sleep disturbances.

5.4.2 Limitations in the Availability and Quality of Noise Exposure Data

It is not uncommon for noise effects researchers to be confronted with noise exposure data scarcity. Precise predictions about the impact of noise on the whole population require the availability of exposure data in all level ranges, meaning also in level ranges below the typical reporting thresholds. For example, the main gap in e.g. the EU environmental noise directive (END) exposure data for health impact assessments is the non-coverage of populations beyond noise hotspots. Within the framework of the END, the mandatory noise level reporting thresholds are, with 55 dB Lden and 50 dB Lnight, well above the limits of harmfulness according to the WHO. It becomes evident that such scarcity not only hampers the estimation of risks at low exposure levels in primary studies, but also affects proper exposure distribution estimates down to low noise levels in entire populations. The true health impacts across whole populations are thus prone to be underestimated. Whereas this problem persists even in the most developed countries, we must realize that many countries outside Europe do not have any estimates of the noise exposure distribution of their population. Many countries simply lack the resources to calculate complete (or indeed any) noise maps. Makeshift solutions are needed to overcome this lack of data. For instance, for low exposure data, a flat distribution in the low noise region can be assumed as a first rough approximation. With the aid of the rule of thumb-conversions, the exposure at night can be approximated from day-related metrics like Lden or Lday (Brink et al. 2018) and so on.

Reliable exposure distribution estimates for larger areas, e.g. for the whole of Europe, are often hampered by incompatibilities of different national noise exposure assessment models and methods. In Europe alone, more than 10 different road traffic calculation models are commonly used currently (European Environment Agency (EEA) 2020).

To carry out large-scale (or large area) environmental noise impact assessments, one can often only use ready-made noise maps that are provided in one or the other format (i.e. raster, polygon or polyline) and with different level ranges and levels of categorization. Very often, the exact distribution of the population per decibel level is not known, as reporting is done in 5 or 2.5 dB wide bins. In the wake of a health impact assessment of road traffic noise in more than 700 European cities and 25 capitals, Khomenko and colleagues came to the conclusion that a large majority of noise maps available for these cities, were of moderate or only low quality (Khomenko et al. 2022). Besides their low quality, it is very likely that noise maps from different countries and generated by different modeling methods cannot really be compared to each other. This assumption is also supported, for example, by the fact that in the aforementioned study, the authors found very large differences in the percentage of people exposed to levels above 55 dB Lden across the 25 European capitals. These percentages ranged from an unrealistic 99.8% in Sofia, 93.3% in Luxembourg city and 81.6% in Riga to 33.8% in London, 30.3% in Brussels and 29.8% in Berlin. Although differences between cities are to be expected to some extent, e.g. due to different residential and urban design patterns, the differences observed here are so large that it seems likely that a large part is due to methodological differences in the exposure assessment, or even to erroneous calculations themselves. Such are of course important sources of uncertainty in large-scale impact assessments. Differences in noise exposure assessment translate directly into the local estimates and thus have a serious influence on the overall result of such exercises. Solid noise impact assessments therefore need more harmonized and comparable noise exposure data in the future.

5.4.3 Defining the Lowest Reference (Threshold) Noise Exposure

One can assume, that the proportion of persons that are exposed to rather low levels of noise is very large, especially in the case of aircraft and railway noise – compared to road traffic noise to which the population is exposed in a normal distribution-like fashion. The choice of a threshold or reference level above which the noise impact assessment is carried out (that means where people or cases are effectively 'counted'), has ipso facto significant implications for the estimated impact. If, say, of 100,000 persons in a city, 10% are exposed to a Lden level of 55 dB and where the percentage highly annoyed (%HA) is 10%, but at 50 dB with 90% of the people the corresponding %HA-figure is 5%, the contribution to the overall effect of the lower exposure group would still be more than twice as large than that of the higher exposure group. Using a reference level higher than 50 dB would then disregard the considerable effect on the population below the reference value. By tendency, the older a study is, the higher the reference levels for the derivation of an exposure – response relationship were assumed. This means that often no effect estimates are available for low to moderate noise levels. Thus, most of the noise health risk assessments so far defined reference levels roughly in the range between 50 and 55 dB (e.g. Khomenko et al. 2022), with some exceptions (e.g. Ecoplan 2019 or Hegewald et al. 2021) which nowadays can probably be regarded as too high.

But how to arrive at a reasonable reference value? A straightforward approach would be to simply use the WHO guideline values, because of their ubiquity. However, these values represent levels, for which the WHO Guideline Development Group was "confident that there is an increased risk of adverse health effects" (WHO 2018, Page 20). This means that even if the guideline values would be respected in a certain exposure scenario, there would already be health effects below these values.

The second approach draws from the lowest levels reported in epidemiological exposure – response studies. The WHO calculated a weighted average of the lowest exposure values (reference values) used in the individual studies included in the respective meta-analyses underpinning their recommendations (in particular (Van Kempen et al. 2018)) and used it as a reference value for the respective source/ endpoint (to be precise: road traffic noise and ischaemic heart disease: 53 dB Lden; aircraft noise and ischaemic heart disease: 47 dB Lden; there was no evidence of sufficient quality for other sources/endpoint combinations). The question here is: do we have good reasons to adopt a certain reference level just because it is or was the lowest one reported or used as the reference category in epidemiological studies? Probably not necessarily, as most noise effect studies for all kinds of outcomes so far found no evidence for a natural threshold below which no noise effects would occur. Indeed, several newer original studies that covered a larger range of trans-portation noise exposure indicate that at least for cardiometabolic endpoints, the exposure response relationships are approximately linear across a large exposure range. Thus, risk estimates (risk increase per decibel increase) might well be extrap-olatable into lower exposure ranges. Adopting such a view, it seems appropriate to set reference values more or less arbitrarily, e.g. at 40 dB Lden, as proposed by Röösli and colleagues (Röösli et al. 2022).

5.4.4 Unstudied or Understudied Environmental Noise Sources

While transportation noise (road, rail, air) and its effects are well studied, exposure – response relationships have not yet, or only to a limited degree, been developed for other types of noise that may be equally harmful to health. By focusing on transportation noise alone, we run the risk of underestimating the burden of noise exposure in general. We should not fail to mention so-called "neighborhood noise" as another significant noise type in everyday life. One reason for the paucity of scientific studies concerning annoyance caused by this type of noise is that it is rather difficult to express in quantitative terms. While there exist studies investigating annoyance and/or self-reported sleep disturbances from such sources as for example frogs croaking (Sasazawa et al. 2002), there is no robust evidence for the effects of non-transportation-related sources on outcomes such as cardiovascular disease or diabetes. This means that additional assumptions about the transferability of exposure – response functions become necessary, if one wants to estimate the health impact of such seldom studied sources. What seems already clear is that a few rather novel, previously uninvestigated noise sources, are becoming increasingly important. These include, for example, heat pumps or wind turbines. The more widespread such sources will become in the future, the more relevant they are potentially for public health.

5.4.5 "Double Counting" Issues

The WHO Environmental Noise Guidelines provided methodological guidance for conducting noise impact assessments in Sect. 5.5 (WHO 2018). They point out that the quantification of impacts for one combination of noise sources, noise exposure indicator and health outcomes may, to some degree, include impacts attributable to another such combination. Consequently, for any particular combination, possible double counting should be considered. Any population may be exposed to multiple different noise sources, associated with the same health outcome. Moreover, exposure to one noise source may lead to multiple health effects (e.g. noise annoyance and cardiovascular disease, or other comorbidities). Estimated impacts of multiple noise sources or multiple health effects should not be added without recognizing that addition will, in most practical circumstances, lead to some overestimation of the true impact.

5.4.6   Applicability of Generalized Exposure – Response Functions in Diverse Settings Around the Globe

At the beginning of this section, we noted that there are many uncertainties regarding the correct quantification of noise impacts on health. One often discussed uncertainty concerns the transferability of exposure – response relationships from locations where studies were carried out or data gathered, to different locations, e.g. countries, and/or contexts. This is particularly the case with noise annoyance, but is probably not uncommon with other health outcomes also. For example, differences in what is perceived as annoying are substantial, even within Europe. Different vehicle mixes, building standards, noise exposure modelling, to name just a few, do their further. Despite this, so-called generalized exposure – response relationships for annoyance (e.g. the seminal "EU curves") are still widely used for noise impact assessment and action planning in the EU, after they became an official tool in 2002 (European Commission 2002). But whether such transfers are really possible without further ado is questionable, because almost no socio-acoustic study conducted so far has been able to confirm the validity of generalized curves in specific local contexts. This can be well demonstrated, for example, in the case of aircraft noise using the Community Tolerance Level (CTL) method, where it becomes apparent that each airport in principle produces its own exposure – response curve (Fidell et al. 2011). Most of the time, "local" survey results are above or below the "generalized" ones, which is an indication that local conditions play a too strong role, so that the application of generalized curves can describe the impact only with a lot of uncertainty. Therefore, the WHO recommends in their guidelines that whenever possible, exposure – response curves derived in a local context should be used to assess the relationship between noise and annoyance in that same local situation. For self-reported outcomes, such as annoyance, the number of surveys available using this approach is large, which allows for considering the most appropriate one. However, in the domain of physical diseases, the scarcity of relevant studies remains problematic. When local data are not available, and this is the rule rather than an exception, the WHO recommends for health impact assessments, to apply general exposure – response relationships, which is the only viable option anyway (WHO 2018).

5.4.7   Impact of Combined Exposure to Several Noise Sources at the Same Time

In principle, it can be assumed that an exposure situation with several noise sources present at the same location should lead to a larger overall environmental noise impact, than would be attributable to just one noise source alone. However, correctly accounting for separate, but additive impacts from several sources is a very delicate problem to solve. Mostly because most if not all, exposure – response relationships in the literature regard single sources only. But this does not mean that the empirically observed exposure situation under which a source-specific exposure – response relationship was derived, did not have exposure from other noise sources, but just that these – even if present – were simply not considered in the modeling exercise. However, noise sources other than the one primarily in focus in an exposure – response model can influence the effect of the noise source in focus, e.g. through acoustic masking effects. For example, it can be assumed that high road traffic noise levels reduce the annoyance caused by aircraft noise.

Important to note is that the selection of particular combinations of noise sources, noise exposure indicators, and health outcomes used to estimate health impacts from combinations of noise sources will depend on the particular policies and/or measures being evaluated.

5.4.8 Lack of Well-Founded Disability Weights (DWs) for Noise

As mentioned previously in this chapter, DWs are needed to determine the non-fatal component of DALYs. Attributing weights to health conditions or health states is not a purely scientific exercise, as this process involves social and individual values and preferences. As the less severe responses to environmental conditions tend to affect the highest number of people, variation in DWs has a large impact on the burden of disease estimates. De Hollander et al. therefore concluded, that DWs for noise effects such as noise annoyance or noise-induced sleep disturbance, must be established with great care (De Hollander et al. 1999). However, regarding the quantification of noise effects on health, DWs are not always easy to handle. While on one hand, the scientific literature is lacking DWs for some of the commonly studied noise effects (like e.g. hypertension, which is mainly viewed as a risk factor and not as a health outcome in itself), for others, e.g., annoyance, the empirical foundations have been considered to be rather weak (Van Kamp et al. 2018). Yet DWs have a strong impact on calculated DALY figures – as mentioned before. For example, of the calculated DALYs in the aforementioned WHO 2011 report (WHO 2011), the largest fraction (roughly 900 000 DALYs) resulted from morbidity due to sleep disturbances. This large number is thus sensitive to the DW chosen (for the 2011 WHO report: 0.07). That DW used was assigned a standard uncertainty of 0.04–0.10 by WHO, which – compared to the more moderate uncertainties in exposure estimation – is quite substantial. In such cases, researchers sometimes opt to derive a DW specifically for use in the study. The problem with this solution is that these bespoke DWs cannot be used together with DWs from other studies if they are derived with different methods. The major reason for this is that the methods that are used to determine the DW have a large impact on its value, resulting in a different use of the 0–1 DW scale. However, for many years, this was the case, e.g. the high annoyance DW. The high annoyance DW that was used to calculate environmental noise DALYs was not based on the same methods as the DWs for the other health outcomes that were included in the DALY calculations.

Very recently, in 2022, a DW measurement study was commissioned by the WHO that derived DWs for a range of health outcomes associated with environmental noise exposure, including, for the first time, high annoyance and severe sleep disturbance (Charalampous et al. 2022, 2024). Use of these new DWs will improve the precision of the calculation of environmental noise DALYs in the future.

5.5 Conclusions and Outlook

Urbanization and economic growth, including the ever increasing rise of traffic on the ground and in the air, are the main drivers of environmental noise and its impact on health worldwide. Policymakers in most countries are faced with increasing public pressure to remedy the noise problems and/or are required by legislation to reduce the unwanted health effects of noise. Because corresponding actions often involve considerable efforts (including financial efforts), it might be desirable to prioritize some environmental mitigation measures over others. For this purpose, it is necessary to know the "size of the problem". Methods referred to as "health impact assessment" can be used to provide answers to that kind of questions and inform public health decisions. This chapter has shown that to estimate the burden of disease due to environmental noise – that means the quantification of noise effects on the population level – a range of techniques and indicators are available, with the disability-adjusted live years (DALY) metric being the most popular. However, it has also become clear that several gaps in the coverage and representativeness of noise exposure and health data and – depending on health effects in focus – the paucity of viable exposure – response relationships must be overcome in the future. Our goal must be to be able to produce better and less biased large-scale health impact assessments of noise effects to better inform policymakers and the public about the true health costs of noise. We are just at the beginning of a long journey. But as of today already, several studies, most of them carried out in Europe, have demonstrated the enormous health loss caused by noise, a result that should encourage politicians and authorities, indeed all of us, to do more to combat noise.

References

Aasvang GM, Stockfelt L, Sorensen M, Turunen AW, Roswall N, Yli-Tuomi T et al (2022) Burden of disease due to transportation noise in the Nordic countries. Glasgow, Internoise

Brink M, Schreckenberg D, Thomann G, Basner M (2010) Aircraft noise indexes for effect oriented noise assessment. Acta Acust Acust 96:1012–1025

Brink M, Schaffer B, Vienneau D, Pieren R, Foraster M, Eze IC et al (2019) Self-reported sleep disturbance from road, rail and aircraft noise: exposure-response relationships and effect modifiers in the SiRENE study. Int J Environ Res Public Health 16(21):4186

Brink M, Schäffer B, Pieren R, Wunderli JM (2018) Conversion between noise exposure indicators Leq24h, LDay, LEvening, LNight, Ldn and Lden: principles and practical guidance. Int J Hyg Environ Health 221(1):54–63

Charalampous P, Polinder S, Wothge J, von der Lippe E, Haagsma JA (2022) A systematic literature review of disability weights measurement studies: evolution of methodological choices. Arch Public Health 80(1):91

Charalampous P, Maas CCHM, Haagsma JA (2024) Disability weights for environmental noise-related health states: results of a disability weights measurement study in Europe. BMJ Public Health 2:e000470. https://doi.org/10.1136/bmjph-2023-000470

Cochrane Collaboration. GRADE Approach (2023) [cited 2023 13 Oct 2023]. https://training. cochrane.org/grade-approach#:~:text=GRADE%20is%20a%20systematic%20approach,rev iews%20and%20other%20evidence%20syntheses

Correia AW, Peters JL, Levy JI, Melly S, Dominici F (2013) Residential exposure to aircraft noise and hospital admissions for cardiovascular diseases: multi-airport retrospective study. BMJ 347:f5561

De Hollander AE, Melse JM, Lebret E, Kramers PG (1999) An aggregate public health indicator to represent the impact of multiple environmental exposures. Epidemiol 10(5):606–617

Ecoplan (2019) Auswirkungen des Verkehrslärms auf die Gesundheit-Berechnung von DALY für die Schweiz-Schlussbericht im Auftrag des Bundesamtes für Umwelt. Bern and Altdorf

European Commission (2002) Position paper on dose response relationships between transportation noise and annoyance: Luxembourg: office for official publications of the European communities. Contract No.:ISBN 92–894–3894–0

European Environment Agency (EEA) (2020) Environmental noise in Europe—2020. Publications Office of the European Union, Luxembourg

Fenech B, Clark S (2022) An update to the WHO 2018 environmental noise guidelines exposure response relationships for annoyance from road and railway noise. Internoise 2021; Glasgow

Fenech B, Janssen S (2023) Review of noise policies and economic evaluations relevant to ICBEN for the period 2021–2022. ICBEN 2023, Belgrade

Fidell S, Mestre V, Schomer P, Berry B, Gjestland T, Vallet M et al (2011) A first-principles model for estimating the prevalence of annoyance with aircraft noise exposure. J Acoust Soc Am 130(2):791–806

Field MJ, Gold MR (1998a) Summarizing population health: directions for the development and application of population metrics. Institute of Medicine: National Academy Press, Washington (DC). Report No.: 0309060990

Field MJ, Gold MR (1998b) Summarising population health: directions for the development and application of population health metrics. National Academy Press, Institute of Medicine, Washington D.C.

Gao T, Wang XC, Chen R, Ngo HH, Guo W (2015) Disability adjusted life year (DALY): a useful tool for quantitative assessment of environmental pollution. Sci Total Environ 511:268–287

GBD 2019 Diseases and Injuries Collaborators (2020) Global burden of 369 diseases and injuries in 204 countries and territories, 1990–2019: a systematic analysis for the global burden of disease study 2019. Lancet 396(10258):1204–1222. https://doi.org/10.1016/S0140-6736(20)30925-9. Erratum in: Lancet 396(10262):1562. PMID: 33069326; PMCID: PMC7567026

Guski R, Schreckenberg D, Schuemer R (2017) WHO environmental noise guidelines for the European region: a systematic review on environmental noise and annoyance. Int J Environ Res Public Health 14(12):1539

Haagsma JA, Polinder S, Cassini A, Colzani E, Havelaar AH (2014) Review of disability weight studies: comparison of methodological choices and values. Popul Health Metr 12:20

Hanninen O, Knol AB, Jantunen M, Lim TA, Conrad A, Rappolder M et al (2014) Environmental burden of disease in Europe: assessing nine risk factors in six countries. Environ Health Perspect 122(5):439–46

Hegewald J, Schubert M, Lochmann M, Seidler A (2021) The burden of disease due to road traffic noise in Hesse, Germany. Int J Environ Res Public Health 18(17)

Joffe M, Mindell J (2005) Health impact assessment. Occup Environ Med 62(12):907–912

Khomenko S, Cirach M, Barrera-Gomez J, Pereira-Barboza E, Iungman T, Mueller N et al (2022) Impact of road traffic noise on annoyance and preventable mortality in European cities: a health impact assessment. Environ Int 162:107160

Klatte M, Bergström K, Lachmann T (2013) Does noise affect learning? A short review on noise effects on cognitive performance in children. Front Psychol 4:578

Knol AB, Petersen AC, van der Sluijs JP, Lebret E (2009) Dealing with uncertainties in environmental burden of disease assessment. Environ Health: Glob Access Sci Source 8:21

Munzel T, Sorensen M, Daiber A (2021) Transportation noise pollution and cardiovascular disease. Nat Rev Cardiol 18(9):619–636

Murray CJL, Lopez AD (eds) (1996) The global burden of disease: a comprehensive assessment of mortality and disability from diseases, injuries and risk factors in 1990 and projected to 2020. Harvard University Press, Cambridge

Murray CJ, Acharya AK (1997) Understanding DALYs (disability-adjusted life years). J Health Econ 16(6):703–730

Murray CJ, Salomon JA, Mathers CD (2002) Summary measures of population health: concepts, ethics, measurement and applications. World Health Organization, Geneva

Murray CJL, Vos T, Lozano R et al (2012) Disability-adjusted life years (DALYs) for 291 diseases and injuries in 21 regions, 1990–2010: a systematic analysis for the global burden of disease study 2010. Lancet 380:2197–2223

Prüss-Üstün et al (2016) Preventing disease through healthy environments

Röösli M, Wunderli JM, Brink M, Cajochen C, Probst-Hensch N (2019) Die SiRENE-studie. Swiss Medical Forum 19(0506):77–82

Röösli M, Wicki B, Vienneau D (2022) Reflections on burden of disease and health impact assessment methods for noise. Internoise, Glasgow

Sasazawa Y, Xin P, Suzuki S, Kawada T, Kuroiwa M, Tamura Y (2002) Different effects of road traffic noise and frogs' croaking on night sleep. J Sound Vib 250(1):91–99

Shamsipour M, Zaredar N, Monazzam MR, Namvar Z, Mohammadpour S (2022) Burden of diseases attributed to traffic noise in the metropolis of Tehran in 2017. Environ Pollut 15(301):119042

Smith MG, Cordoza M, Basner M (2022) Environmental noise and effects on sleep: an update to the WHO systematic review and meta-analysis. Environ Health Perspect 130(7):76001

Sorensen M (2023) Emerging adverse health effects of traffic noise. ICBEN 2023, Belgrade

Thacher JD, Hvidtfeldt UA, Poulsen AH, Raaschou-Nielsen O, Ketzel M, Brandt J et al (2020) Long-term residential road traffic noise and mortality in a Danish cohort. Environ Res 187:109633

Van Kamp I, Schreckenberg D, van Kempen E, Basner M, Brown A, Clark C et al (2018) Study on methodology to perform environmental noise and health assessment. RIVM 2018–012

Van Kempen E, Casas M, Pershagen G, Foraster M (2018) WHO environmental noise guidelines for the European region: a systematic review on environmental noise and cardiovascular and metabolic effects: a summary. Int J Environ Res Public Health 15(2)

Veber T, Tamm T, Rundva M, Kriit HK, Pyko A, Orru H (2022) Health impact assessment of transportation noise in two Estonian cities. Environ Res 204(Pt C):112319

Vienneau D, Saucy A, Schaffer B, Fluckiger B, Tangermann L, Stafoggia M et al (2022) Transportation noise exposure and cardiovascular mortality: 15-years of follow-up in a nationwide prospective cohort in Switzerland. Environ Int 158:106974

WHO (2011) Burden of disease from environmental noise. Quantification of healthy life years lost in Europe. https://www.euro.who.int/en/health-topics/environment-and-health/noise/publicati ons/2011/burden-of-disease-from-environmental-noise.-quantification-of-healthy-life-years-lost-in-europe

WHO (2012) Methodological guidance for estimating the burden of diseasse for environmental noise
WHO (2018) Environmental Noise Guidelines for the European Region. http://www.euro.who.int/en/health-topics/environment-and-health/noise/publications/2018/environmental-noise-guidelines-for-the-european-region-2018
Worldbank (1993) World development report 1993: investing in health. Oxford University Press, New York, p 1993

Chapter 6
Economics and Environmental Noise

Ronny Klæboe

> *The task is not just to understand the world but to change it.*
> *—Karl Marx*

6.1 Introduction

The previous chapters of this book have shown that noise can be understood as unwanted and/or harmful sound (Chap. 3), that noise can have considerable negative health effects (Chap. 4) and that the burdens of these health effects are considerable (Chap. 5). Rainer Guski in Chap. 3 explains that the meaning of sound is context dependent.

This chapter focuses on quantifying the costs of environmental noise in monetary terms (monetisation) that can be used for taxation of travel and freight activities producing unwanted noise and other externalities. Monetisation also allows acoustic consequences to be included in economic analyses of infrastructure investments, efforts to reduce noise and efforts to improve the acoustic environment. It is when the external costs of noise are used for taxation and the economic benefits of an improved acoustic environment are included in the planning of transport and building infrastructure, urban and regional development processes that they affect policies and decision-making. Finally, we mention some of the shortcomings of current approaches and the scope for further research and caveats.

R. Klæboe (✉)
Institute of Transport Economics, Oslo, Norway
e-mail: ronny@klaeboe.info

6.2   Costs of Environmental Noise

Human impacts of noise are dominated by noise from transport sources (routes and terminals), and we will focus on these sources. As a result of the increasingly strict Euro vehicle emission standard regimes, local emissions of combustion particles and exhaust gases from private and commercial vehicles have been greatly reduced over the last decades despite traffic increases. However, noise impacts have not diminished, and have also in many countries such as Norway (Engelien and Steinnes 2021), increased over time. The improvements from action plans and noise reduction measures have not managed to counteract the impact of increasing traffic and more people located close to and becoming affected by urban roads (urbanisation).

One reason for the persistence of traffic noise impacts is that most of the noise from road traffic is not from the engines but is produced by tyre/road surface interaction at higher speeds. Stricter regulations of noise emissions from new vehicles at lower speeds and electrification reduce noise emissions at lower speeds. But on motorways and major thoroughfares with many vehicles and high speeds, it is the interaction between the road surfaces and tires that dominates.

The noise from heavy vehicle engines still plays a role at lower speeds and noise from powertrains with tonal components can travel far. Most of the rail traffic noise and vibrations are also due to the interaction between train wheels and rail surfaces. At lower speeds noise from locomotive engines and at high speeds aerodynamic noise play a larger role.

Other major noise sources include industrial sites and service centres, noise from shooting ranges and military training facilities, concerts, sports, and motor events, building construction and repair and human outdoor activities such as playing and partying. Figure 6.1 provides an overview of the noise from the most important Norwegian environmental noise sources and the number of citizens estimated being exposed to them.

Environmental noise affects residential areas, workplaces, recreational areas, city centres, public places and outdoor areas, footpaths, and lanes for cycling. In addition to the health impact costs, noise disturbances and interferences reduce life quality and affect social relations. People avoid and/or reduce time spent in areas impacted by noise and properties and neighbourhoods affected by noise lose property value (Nelson 2008). There are large costs associated with public and private efforts to deter noisy activities and reduce noise production. Low-noise road surfaces are worn down and damaged and need to be resurfaced and maintained.

There are also significant opportunity losses due to zoning restrictions, speed limits, night-time restrictions and of limiting access to noisy neighbourhoods. Costly private and public measures are employed for noise shielding and barriers, hampering the propagation of noise, improving noise insulation and protecting individuals against noise along and around noisy routes, sites, and places.

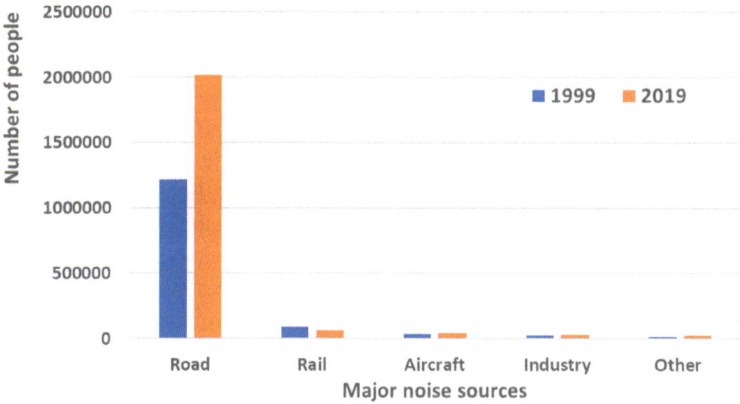

Fig. 6.1 Estimate of the number of people affected by noise (L_{den}) above 55 dB from different sources in Norway 1999 (population 4,462 million) and 2019 (population 5,348 million) (Engelien and Steinnes 2021)

6.3 Noise Impact Costs and Taxation

A basic premise in social economics is that transport users, activity centres, service providers and other noise producers only consider their own costs but neglect the direct and indirect costs their activities impose on others. Such external costs include the damage costs of noise, local and global emissions, costs to others of accidents, road wear and the part of queuing costs that are imposed on others. Roads with high traffic volumes, high vehicle speeds, and heavy vehicles and buses are perceived as imposing and intimidating, as well as constituting barriers limiting neighbourhood activities and movements. Road and rail traffic cause safety risks and insecurity for the population located along or nearby roads and rail tracks.

Not charging producers and site owners for these external costs, means trip-making becomes too cheap, thereby motivating users to take more and longer trips than is socio-economically optimal. Drivers also make route choices without proper consideration for the impacts on neighbourhoods, contextual factors, critical blockage points and time slots (rush hours, start and end of sports and cultural events, school starting and closing hours).

To avoid excessive damages, authorities may use taxation to reduce noise at the source. A well-known taxation instrument is a Pigouvian tax (Pigou 1924) recommending that the transport user should be taxed a sum equal to the marginal damage costs that their activity gives rise to. These costs thereby become internalized and are no longer neglected. To find the appropriate level of taxation, a recent effort by Norwegian transport authorities assesses the external costs of transport (Rødseth et al. 2019). In these calculations, health impact costs for the first time were derived from DALYs and we describe these calculations for Norway as an example of how to proceed. In addition to the health impacts, the life quality impact cost of being moderately annoyed was included.

6.4   Increases in Noise Exposure from Traffic Increases

For taxation the marginal costs are of interest; the average costs associated with an extra vehicle travelling on existing roads. Environmental noise health and life-quality impacts are important parts of these costs. The change in noise exposure now and in the future is a function of the changes in traffic flows, based on the existing and future vehicle fleets, speeds, bottlenecks and queuing, spatial layouts, and distances between each road or rail segments and each of the groups of citizens affected by their traffic flows. To calculate the increase in noise level resulting from an increase in traffic at a national level it is necessary to establish an inventory of all road stretches (state, county and municipality) with the number of light and heavy vehicles along each of the road segments, their average speeds, and the number of people affected at different distances from each of the road segments. Terrain, noise screens and intervening buildings need to be considered. For rail traffic, it is necessary to know the types of locomotives used on different stretches along with the length and speeds of trains and distances travelled in different time periods. For air traffic the traffic volumes of the airports and aircraft types are of interest and how much noise an extra flight generates given runway usages, time slots, and departure and landing patterns.

The relative increase in the noise level in dB of adding a certain number of vehicles on a road is smaller if there already are many vehicles. Doubling the number of vehicles on a road with 5000 vehicles per day results in an increase of $+3$ dB, whereas the same number of additional vehicles only generates $+1$ dB when added to a road with 20,000 vehicles. On the other hand, major roads often run through more densely populated areas where noise increases affect more people. With less traffic, for instance at night, the relative increase in noise level of an extra vehicle will be higher.

When noise emissions increase the area of influence increases. For a line source such as noise emissions along a busy road or rail stretch, each increase of 3 dB means that the potential noise impact area doubles. Calculations based on the average distances between dwellings and line sources in Norway resulted in the addition of 10% to the noise cost to take care of this change in the influence area.

6.5   Health Impact Costs of Environmental Noise

In a study for the Norwegian infrastructure authorities (Rødseth et al. 2019), the marginal costs of many long- and short-term health consequences of road, rail and air transport needed to be monetized so they could be added to the external costs of greenhouse gases, road wear, barriers and queuing.

Lacking specific valuations of each of the health consequences, it was decided to apply a unit cost to each DALY. There was no previous estimate of this unit cost, and an estimate was therefore obtained indirectly from the value of a statistical life used in Norwegian traffic safety research. Since a person lost in a traffic accident in Norway on average had an estimated 37 remaining years to live, the value of 1 DALY was calculated as the annuity that over 37 years is equal in sum to the value of a statistical life lost. This resulted in the value of 1,611,000 NOK or approximately € 161,100 (2019 values) which was used in the calculations of costs of the different health effects.

6.5.1   Cost of Becoming Highly Annoyed

The disability weight for high noise annoyance differs between studies. The Norwegian study (Rødseth et al. 2019) used a value of 0.02. This means that being highly annoyed for 50 years is set equivalent to the loss of 1 life year. Using € 161,100 for 1 DALY, the cost per person of being highly noise annoyed for 1 year is € 3220 (0.02 × € 161,100).

To find the average high annoyance costs caused by a noise increase of 1 dB, we need to know the increase in the number of citizens thereby becoming highly annoyed. The estimated parameters of a grouped regression analysis (Miedema and Oudshoorn 2001) were used to calculate the proportion that are highly annoyed on a modern 5-point (see Fig. 6.2) rather than the older 7-point annoyance scale (Schultz 1978) as a function of L_{den}. The slopes of each of these curves indicate the increased proportion of the population becoming affected by a 1 dB increase given their initial exposure level.

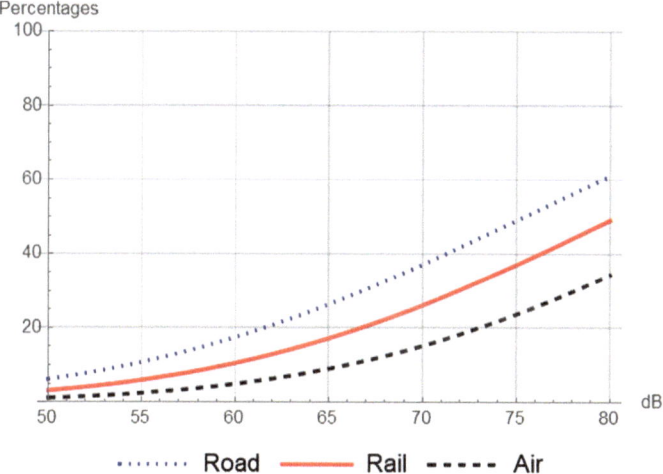

Fig. 6.2 Impact curves for persons highly annoyed (%HA) as a function of noise exposure (L_{den}). Based on grouped regression parameters (Miedema and Oudshoorn 2001)

Table 6.1 Marginal costs per year per person highly annoyed from an increase in noise exposure (L_{den}) of 1 dB

Road traffic	€2019	Rail traffic	€2019	Air traffic	€2019
52–65 dB	33.40	54–65 dB	21.30	54–65 dB	43.50
65–80 dB	75.40	65–80 dB	59.90	65–80 dB	74.40

For calculations see Text Box: 'Details of the Marginal Cost Calculations'

We have extracted the main features of these curves by using piecewise linear interpolation, multiplied by the cost of being highly annoyed and incorporated the increase in influence area to find the marginal costs (see Table 6.1).

This simplification makes it easier for practitioners to apply results (see Table 6.6).

6.5.2 Cost of Being Highly Sleep Disturbed

The disability weight for high sleep disturbance varies between studies. The disability weight of 0.07 was used in the Norwegian study. This means that being highly sleep disturbed for 14.3 years is set equivalent to the loss of 1 life year. With the value of € 161,100 for 1 DALY, the cost per person of being highly noise annoyed for 1 year is $0.07 \times$ € $161,100 =$ € 1123.

A recent meta-analysis of various studies (Basner and McGuire 2018) describes the probability of being highly sleep disturbed (%HSD) as a function of L_{night} using second-degree polynomials (see Fig. 6.3).

Road %HSD $= 19.4312 - 0.9336\, L_{night} + 0.0126\, L_{night}^2$

Train %HSD $= 67.5406 - 3.1852\, L_{night} + 0.0391\, L_{night}^2$

Aircraft %HSD $= 16.7885 - 0.9293\, L_{night} + 0.0198\, L_{night}^2$

L_{night} is the equivalent noise exposure on the most exposed façade at night.

Approximating the curves using piecewise linear interpolation and incorporating the increased influence area), we obtain the marginal cost in Table 6.2.

Many tools and calculation methods do not provide separate L_{night} values. However, L_{night} values can be approximated with $L_{den} -6$ for road traffic (Basner and McGuire 2018) and L_{night} can in Norway be approximated with $L_{den} -7$ for rail traffic (Rødseth et al. 2019) and we thus obtain the results in Table 6.3.

Due to night restrictions and different activity patterns for aircraft at different airports, we do not have a suitable general conversion factor and require an estimate of L_{night} before including the costs of sleep disturbances for those with $L_{night} > 50$ dB.

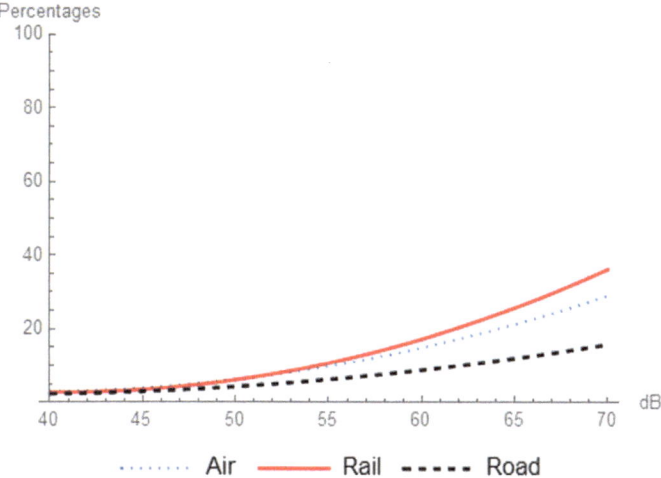

Fig. 6.3 Impact curves for persons who are highly disturbed by sleep as a function of the equivalent noise level at night (L_{night}) on the most exposed façade. Based on Basner and McGuire (Basner and McGuire 2018)

Table 6.2 Marginal cost of becoming highly sleep disturbed per person per dB (L_{night})

L_{night}	€2019
Road traffic > 50 dB	71.70
Rail traffic > 50 dB	186.90
Air traffic > 50 dB	141.90

For calculations see Text Box: 'Details of the Marginal Cost Calculations'

Table 6.3 Marginal cost of becoming highly sleep disturbed per person per dB (L_{den})

L_{den}	€2019
Road traffic > 56 dB	71.70
Rail traffic > 57 dB	186.90
Air traffic L_{night} > 50 dB	141.90

6.5.3   Cost of Ischemic Cardiovascular Disease (Road Traffic Only)

The cost for people who die from, or contract ischemic heart disease is equivalent to the average number of years lost. In the Global Burden of Disease (Øverland et al. 2016) this was calculated to be 11.376 years. With the value of € 161,100 for 1 DALY, the cost per person of death from ischemic heart disease in Norway is 11.376 * € 161,100 = € 1,832,673.60.

Fig. 6.4 Relative risk (RR) for contracting ischemic heart disease as a function of exposure (L_{den}). Based on Van Kempen et al. (Van Kempen et al. 2018)

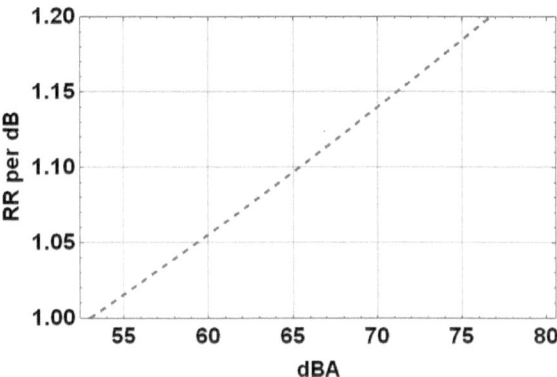

Table 6.4 Marginal cost of ischemic heart disease per person per dB (L_{den})

L_{den}	€$_{2019}$
Road traffic > 53 dB	12.10

For calculations see Text Box: 'Details of the Marginal Cost Calculations'

Van Kempen et al. (Van Kempen et al. 2018) indicate an increase in the risk of contracting an ischemic cardiovascular disease of 8% per 10 dB increase in noise exposure (L_{den}). The relative risk of a smaller change of 1 dB can be approximated by $1.08^{\frac{1}{10}} = 1.00773$. With a cut point of 53 dB, the relative risk of contracting an ischemic cardiovascular heart disease at a given noise level x then becomes $RR(x) = 1.00773 * 10^{(53-x)}$ (see Fig. 6.4).

In Norway, the population risk for ischemic heart disease was 0.0007568 in 2016. This number includes both incidences in areas affected by different levels of noise, and in quiet areas.

Given the relationship between exposure and risk, it is possible to calculate backwards within each exposure interval and accumulate results to derive the population baseline risk of myocardial heart disease without noise. Given the baseline risk, we can subsequently find the marginal cost of a noise increase of 1 dB also considering the increase in influence area (see Table 6.4).

6.6   The Life Quality Cost of Becoming Moderately Annoyed

When undertaking economic analyses of projects that are not limited to the health sector, all relevant adverse effects should be considered. Willingness to pay and hedonic pricing studies clearly show that people attach a value to being moderately annoyed, and the Life Quality impairment cost of moderate annoyance should consequently be included as part of the environmental cost of noise. The practical consequence is that the large number of citizens affected by moderate levels of noise are not neglected during planning. Using results from an older willingness to pay study (Thune-Larsen et al. 2014) and adjusting for growth in wages since the study was undertaken we get a cost of € 397 per dB per person moderately annoyed when taking into account the increased area of influence for road and rail.

From Fig. 6.2, we can find how many people are a little and/or moderately annoyed by taking the difference between the cumulative impact curves for those who are slightly and/or moderately annoyed and those that are highly annoyed. For resulting relationships (see Fig. 6.5).

Using piecewise linear interpolation to simplify and taking the increase in influence area into account we obtain Table 6.5.

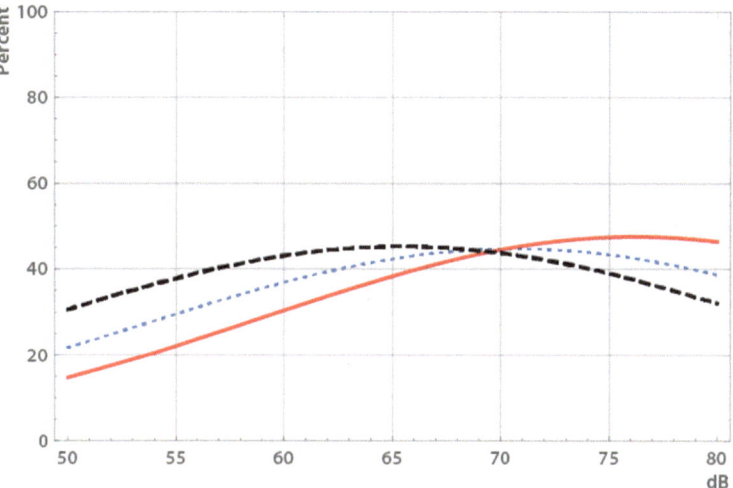

Fig. 6.5 Impact curves for people moderately and slightly annoyed (Between 28 and 72% annoyed). Based on Miedema and Oudshoorn (Miedema and Oudshoorn 2001)

Table 6.5 Marginal cost increase of becoming moderately annoyed

Source		€$_{2019}$
Road traffic noise	$L_{den} > 50$ dB	5.00
Rail traffic noise	$L_{den} > 53$ dB	6.60
Aircraft noise	L_{den} 50–65 dB	3.90
Aircraft noise	L_{den} 65+ dB	−3.50

For calculations see Text Box: 'Details of the Marginal Cost Calculations'

Table 6.6 Health and life quality impact costs of a marginal increase of 1 dB

Source	Exposure			
Road traffic (L_{den})	52 dB	53–55 dB	56–65 dB	65+ dB
Highly annoyed	33.40	33.40	33.40	75.40
Moderately annoyed	5.00	5.00	5.00	5.00
Highly sleep disturbed			71.70	71.70
Ischemic heart disease		12.10	12.10	12.10
Sum	**38.40**	**50.50**	**122.20**	**164.20**
Rail traffic (L_{den})		53–56 dB	56–65 dB	65+ dB
Highly annoyed		21.30	21.30	59.9
Moderately annoyed		6.60	6.60	6.60
Highly sleep disturbed			186.90	186.90
Sum		**27.90**	**214.80**	**253.40**
Air traffic (L_{den})		50–65 dB		65+ dB
Highly annoyed		43.50		74.40
Moderately annoyed		3.90		−3.50
Sum no sleep disturbance	$L_{night} < 50$ dB	**47.40**		**70.90**
Highly sleep disturbed	$L_{night} >= 50$ dB	14.19		14.19
Sum highly sleep disturbed	$L_{night} >= 50$ dB	**61.59**		**85.09**

6.7 Total Health and Life Quality Costs

The previous sections detail the costs of an increase per dB for the most used health and life quality impacts of noise in residential settings. Table 6.6 gives a summary of these costs in € per marginal noise increase (L_{den}, L_{night}) of a 1 dB noise for road, rail and air traffic.

6.8 Economic Analyses of Infrastructure Investments

The major decisions affecting noise impacts on European citizens are undertaken by surface transport and airport authorities in the strategical and tactical planning of new roads, motorways, new rail connections, new airports and landing strips, reconstructing, enlarging and improving existing roads and motorways, and by city authorities and urban planners in developing new city areas and reshaping old.

Infrastructure planning takes place in stages (Bendtsen et al. 2015). The decisions in early planning stages can be heavily influenced by various political, cultural, administrative, and regional group interests, and main routes, sites and building locations may become fixed well before the more detailed acoustic planning starts. To influence decisions at the early planning stages, those affected by the solutions need to become involved and gain support from different environmental health specialists to be able to effect changes. It is better to avoid bad decisions than struggle to overcome the limitations they impose. At the final stages where the major alternatives have been delineated, it is time for more comprehensive analyses.

6.8.1   Social Cost–Benefit Analysis

Socio-economic cost–benefit analysis (CBA) is a comprehensive effort to assess and weigh together all cost and benefit components associated with public investments and maintenance spendings (Mishan 1988). The objective is to achieve the maximum benefit at the lowest cost. The goal is cost **efficiency**, whether society is better off with the measure or not. It differs from cost–benefit analysis undertaken by a specific private or public entity in that it focuses on benefits and costs for society as a whole.

CBA is based on "welfare economics", where all members of society count and where "welfare" is measured in Euro. According to the principle of consumer sovereignty, it is the consumers' own assessments that should serve as the basis for valuations, and not expert opinions or political decisions. Often willingness to pay or hedonic pricing studies are used to establish people´s preferences.

6.8.2   Processing Steps Used in a CBA

We exemplify using elements from CBAs of road transport investments. The objectives can also be to reduce energy consumption and/or production of greenhouse gases or how to avoid or reduce potentially adverse events (natural hazards, pandemics, disruptions). However, the steps are the same (see Fig. 6.6).

The departure point is a set of scenarios where the project participants specify what they want to achieve.

Step 1 of the CBA is to specify one or more sets of measures whereby the desired objectives could potentially be achieved and make an inventory of the components needed together with a list of required work and machine hours (e.g. bulldozers, asphalting column, transport of modules and materials, etc.). Consequences for third parties, such as length and frequency of traffic disruptions on critical links and rerouting may be substantial, and require warning and command vehicles for regulating traffic.

Step 2 is to find the costs of each of these input factors and of proposed upkeep and maintenance policies. This finalizes the cost side.

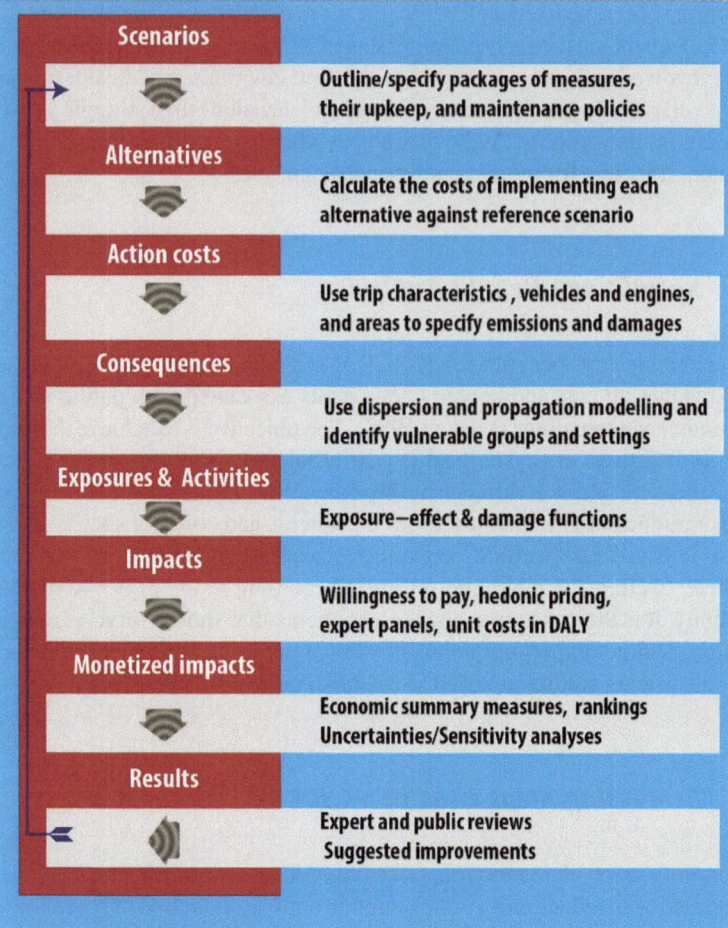

Fig. 6.6 Eight processing steps involved in a CBA

Step 3 is the calculation of the effects based on changes in trip lengths (L/km, μg/km, g/km, accidents/km or kwh/km). The emission amounts and number of accidents depend on the fraction of light and heavy vehicles, the age profiles of the fleets, degree of electrification, road class, speed limits, terrain, climate, etc.

Step 4 requires dispersion and propagation modelling to find the exposures to local air pollution and noise at different distances from the road, taking prevailing winds, screens, terrain and ground conditions into consideration.

Step 5 uses exposure–effect relationships to determine the different health outcomes (short and long-term impacts of air pollution, moderate and severe accidents depending on vehicle types involved, greenhouse gas emissions, road wear, etc.).

Step 6 converts impacts to €, using available ("official") unit costs, hedonic pricing, or willingness to pay results. The effects of interventions (positive and/or negative) measured in € constitute the benefit side of CBA. When a measure has a smaller adverse impact than the reference alternative, the difference (benefit) becomes a positive number.

Step 7 compares the benefits and costs of each intervention option. Results are presented in the appropriate metrics, normally a benefit–cost ratio (BCR) identifying alternatives where the total benefit is greater than the total cost of intervention and ranking them.

We also need to know and communicate the robustness of the results and undertake a sensitivity analysis of the influence of the choice of input factors on the estimated benefits and costs.

Step 8 recommends the best cost-efficient interventions that have positive net benefit and that are robust. However, the overall results hide the fact that some groups may benefit more than others and that some may suffer injustice. At this stage, it is necessary to discuss how the benefits and costs are distributed among the different participant actors, and if necessary, introduce compensatory mechanisms resulting in more balanced projects.

As it is often very costly to reduce noise in established urban areas one should also consider compensation in the form of non-acoustic community initiatives providing other types of health and life quality benefits.

6.8.3   CBA for Selecting Road Surfaces

At EU-wide and national levels, the expenditures on new road pavements, resurfacing old and maintaining road surfaces are very high. The quality of the road surfaces and their noise and other properties are important. The noise from the interaction between the vehicle tyres and the road surface dominates already from quite low speeds. More frequent road pavement resurfacing, using quieter single or multi-layer low-noise asphalts and maintaining their properties over time can be used to reduce noise impacts and reduce Europe-wide annual health impact costs.

How to balance increased costs against improved properties of different road surfaces or choosing a different approach such as making use of noise screens and barriers to reduce noise, is a suitable challenge for undertaking cost–benefit analyses.

It is important that roads with shielding structures are constructed and maintained so that they minimize the noise produced. However, durability, rolling resistance, friction, local emissions to air and water, splashes, tendency for porous surfaces to clogging are also important and these impacts should also be considered in CBAs.

6.8.4   Life Cycle Calculations Can Be of Use

Life cycle analyses (LCA) consider the reduced costs from salvaging materials at the end of projects and reduced resurfacing and maintenance costs by using local material sources, higher quality, and more durable materials over the whole period under consideration. This can be used as inputs to CBAs to include cost savings.

CBAs have the advantage of also looking closely at the benefit side. If dwellings are predominantly located on one side of a road, the cost when using noise screens would be halved. However, the cost of using low-noise asphalt would be the same. The optimal choice of pavements can be to ignore noise production parameters when there are none or relatively few dwellings affected. One could use better quality materials despite higher costs and/or greenhouse gas emissions at locations where many would benefit from the reduced noise production. Both costs and benefits depend on future traffic, future population sizes, and future valuations. CBAs have an advantage in that they consider costs and benefits over the whole project period.

6.8.5   CBAs Deal with Different Time Frames

Cost–benefit analyses are a type of "time machines"; calculators that adjust for the timing of future expenditures and benefits. This makes it possible to compare measures having different time profiles, for example, lower-quality solutions needing frequent and costly maintenance against higher quality alternatives that cost more upfront but promise less frequent and less costly maintenance efforts. If we select a less durable pavement alternative, the number of times within the project period it is necessary to resurface will increase.

Grinding rail surfaces and keeping train wheels well maintained without blemishes and reducing rolling noise and vibrations are significant recurring costs. How often should this be done?

Some costs and benefits are one-time expenditures, and some, such as improved health due to reduced exposure to noise are recurring. To find out which alternative is best it is necessary to harmonize the different streams of costs and benefits. A one-time investment paid upfront can be converted into an annuity by dividing it with an annuity factor. The net present value of costs occurring each 10th year can be summed and thereafter converted to an annuity.

By using a common project horizon to assess different alternatives, a common discount rate, and methodology, the time factor is taken out of the equation. It then becomes possible to rank measures according to their overall economic performance.

6.8.6   CBA of Measures to Improve the Acoustic Environment

It is only at the later more detailed planning stages where specific measures and initiatives to deal specifically with the acoustic environment are undertaken. At his planning stage, it is possible to use more dedicated CBAs to choose between different packages of measures to reduce noise and improve the acoustic environment. Green facades and roofs, soft ground with vegetation may have the potential to not only reduce noise, but also be considered aesthetically pleasing, provide some CO_2 reduction, bind water, contribute to thermal comfort, and bind air-polluting particles.

An advantage of a broader scope taking environmental health and life quality impacts is that policies and measures that achieve multiple goals may become cost-efficient even though noise impact improvements by themselves are not sufficient for implementing the measure.

6.8.7   Comparing Projects Using Benefit–Cost Ratios (BCR)

When undertaking CBAs of many competing alternative solutions, we can compare them using the Benefit–Cost Ratio (BCR). We obtain the BCR by simply dividing the accumulated value of all benefits by the accumulated cost of achieving them. BCR as a relative economic indicator can be considered good for comparing measures/projects of different sizes, but there are also other economic summary indicators such as the net present value (NPV) or the internal rate of return (IRR) that could be utilized.

If the accumulated value of all benefits exceeds the accumulated cost of achieving them, the BCR exceeds one (BCR > 1) and the project is economically "efficient." When more than one project is being evaluated, they can be ranked in order of socio-economic profitability using the BCR. The project with the greatest BCR will be considered for implementation first. However, since there are other projects also waiting for public funding, the benefits should outweigh the costs by a factor of two or more (BCR > 2). These projects are deemed robustly efficient.

When looking at innovative noise abatement solutions in the HOSANNA project (Nilsson et al. 2014) the reasons for having good economic performance were that lattice barriers or tree belts were inexpensive, that the acoustic benefits of porous asphalts more than compensated for the additional costs and that measures such as vegetated façades, provided additional non-acoustic amenity and aesthetic benefits. Some measures, such as tree belts, were both inexpensive and provided such additional benefits. However, all results are project-dependent and depend on how many people benefit from the initiatives and contextual factors.

6.8.8   Cost-Effectiveness Analyses (CEA)

To allocate resources and prioritize measures it can be sufficient to undertake cost-effectiveness analyses (CEA). The objective is not to maximise profit, but to minimize the cost of achieving a predefined acoustic goal. An advantage of a CEA is that there is no need to assign a monetary value for attaining the acoustic or environmental target. The "goods" provided are instead measured in "natural units," such as the reduction in dB, number of units that no longer are in violation of indoor or outdoor environmental limits or guidelines, the reduction in the number of people highly annoyed, etc. Consequently, CEA can be applied in situations where the monetary value of noise impacts has not yet been established. This is currently the case for the acoustic improvement of most outdoor and non-residential situations.

6.8.9   CEA Versus CBA in a Legislative Context

CEAs are often employed in situations in which a mandatory environmental limit needs to be reached or where a political/administrative decision has been made to attain a given environmental improvement. Using leaner products, improved use of materials, recycling, appropriate maintenance, longer life cycles and profiting from "beneficial side effects" mean that more people may benefit from scarce resources available for improving noisy urban environments. Measures having a more efficient design, are more durable, need less maintenance, or employ fewer or cheaper materials, etc., come out on top. If there is a fixed budget, we may seek the solutions that for the given amount result in the maximum number of units satisfying the requirement.

However, reducing noise by enforcing stricter regulation can be very costly or economically inefficient, even when the best solutions have been selected by CEA. A significant disadvantage is that the new legal limits must be implemented regardless of the size of benefits and regardless of the cost. The result of a legislative approach when not warranted is that a lot of money is used to achieve relatively small improvements.

Implementing improvements primarily in situations where the context is favourable, or where many people benefit from the same set of improvements and not everywhere, means that the projects will provide a higher benefit–cost ratio and that more people could potentially benefit from the efforts. This should be the favoured approach when not overruled by social justice concerns or the need to enforce minimum standards.

In some situations, it could be worthwhile to seek mixed policy solutions that both satisfy legislative limits and balance costs and benefits (Klæboe et al. 2011). However, this requires available public funds.

6.9 Shortcomings and Future Challenges

Many impacts of noise are poorly understood or have been subjected to too little research and are consequently not considered when accumulating adverse impacts. Ideally, we should include all the direct and indirect costs of noise disturbances causing inferior cognitive performance, concentration difficulties, enjoyment losses due to reductions in listening and media quality, and degradation of social interactions due to not venturing into noisy areas and inferior quality of human interactions in noisy environments, etc.

A Swedish study shows (see Fig. 6.7) that almost 50% of the population keep their windows closed at night when the noise outside the bedroom window was above 52 dB at night (Öhrström et al. 2006).

When people keep windows and doors closed to avoid the noise, they are less bothered, but at the same time have an adaptation cost. Contact with activities outside the home decreases and ventilation deteriorates, leading to problems with indoor climate. These losses are not currently captured in external noise impact cost assessments.

6.9.1 Acoustic Quality is Important

Traditional health and social policy prioritise those who are most affected and vulnerable groups. The environmental guidelines and limits are usually in the form of minimum requirements, and maximum allowable exceedances. However, cities are increasingly aware that they not only need to avoid environmental ghettos, but also need to pay attention to the qualities of the urban areas to be competitive, attracting businesses and a highly qualified workforce, and design neighbourhoods where it is both good to live and bring up children. Here, the overall quality plays a role. The

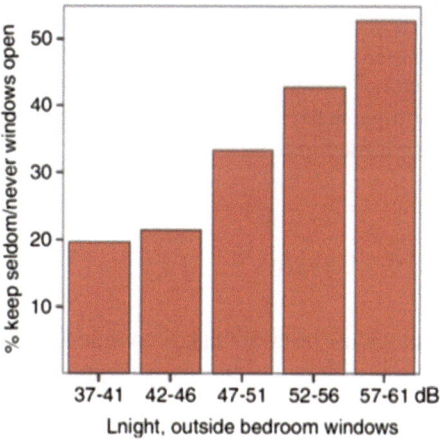

Fig. 6.7 People who seldom or never keep windows open at night because of road traffic noise (Öhrström et al. 2006)

quality of the soundscape should be considered when developing the urban landscape and the blue and green areas of the city (Van Kamp et al. 2015).

Accessible higher quality indoor and outdoor soundscapes can be regarded as local public goods. Public goods, including soundscapes, have economic value, potentially measurable in monetary terms, even if we do not pay directly for their use. However, the monetary value of a public good is not absolute, but relative to its scarcity or abundance and the ease with which it can be replaced with an alternative.

There are several challenges in capturing these effects in addition to the challenges of obtaining good acoustic indicators. The importance of, for example, a quiet area with the potential to provide restoration will thus depend on how many other parks there are in the vicinity and whether the same need could be served or fulfilled by quiet rooms, travelling to more distant recreational areas, or meditation in a suitable indoor environment. The loss of a quiet park is more problematic if no substitutes exist.

The external cost of traffic safety and local air pollution are diminishing due to improvements in passive protection of private and professional vehicles and stricter safety regulation, together with the reductions in local air pollutants from combustion engines due to improved engine technology and electrification. This means that the relative importance of noise has increased. Since noise is easy to produce and very hard to combat, we need to prioritise noise impacts higher than before.

6.9.2   Improving Taxation Schemes

Taxation schemes often use static fuel consumption or distance-based general taxes that reflect the average accumulated marginal costs. This is often in terms of an average per litre fuel, or per km driven within a year. However, the actual time of day, location, number of citizens affected along the route, concurrent events, and queuing conditions for the trip under consideration may differ a lot from average.

A road price differentiated in real time according to real damage costs would socio-economically be a better form of pricing negative externalities in the transport sector. This requires that GIS and location technology become uniformly available for mapping where and when trips are made so that the environmental charges can be adjusted in real time.

6.10   Caveats

One should be aware that economic analyses integrate results from diverse fields, are used in different countries, and apply different methods. Since valuations depend on political frameworks, culture, economic capabilities, the same impacts will be valued differently in different countries and by members of different affluence groups.

There are value-transfer challenges, ethical challenges and practical challenges in bringing everything together. The value of a statistical life varies between countries and different estimation methods may give different values. All input values are also associated with uncertainties.

Sounds, including sounds which many people would call noise, are also not uniformly regarded as negative, since high sound levels in real and/or virtual spheres are associated with industrial or commercial sites generating profit or that are regarded as beneficial, exciting, or even exhilarating, are places where things happen, and where important activities and actors are co-located.

6.11 Hedonic Pricing and Willingness to Pay

The hedonic pricing method is often used to assess the monetary value of local public goods, like noise/soundscape quality. In the hedonic pricing approach, the price differential between dwellings/apartments that are bought and sold on the housing market is analysed taking into account all housing characteristics that are likely to impact the selling price (size, number of bath- rooms, etc.), in addition to neighbour-hood characteristics such as noise, air pollution, vibrations, availability of public services, availability of green and quiet areas.

Statistical techniques can be used to extract the relative importance of, for example, acoustical quality and aesthetics for the valuations. These techniques are however dependent on the availability of suitable indicators.

An alternative economic assessment to hedonic pricing is willingness to pay and stated preference approaches, asking people how much they value different aspects of their environment or presenting choice alternatives where environmental quality levels differ between the alternatives. The willingness to pay method makes it easier to extract valuations of particular aspects (Arsenio et al. 2006).

One could also elicit the respondents' use of municipal or state funds for increased or decreased availability of restorative areas, changes or size of entrance fees deemed acceptable. The extracted values are often given as population averages. When applying the values, it may be useful to consider subpopulations and contextual factors.

Both hedonic pricing and stated preference approaches focus on the valuation of goods and environmental qualities of the actual users, laypeople, and not on the value assigned by architects, landscape planners, psychologists or health professionals. Valuations may be asymmetric in that people may value avoiding a loss higher than profiting by the same amount.

Table 6.7 Marginal costs per person per year highly annoyed from an increase in noise exposure (L_{den}) of 1 dB

Road traffic	Equation	€2019
52–65 dB	0.942717% × € 161,100 × 0.02 × 110%	33.40
65–80 dB	2.12606% × € 161,100 × 0.02 × 110%	75.40
Rail traffic		
54–65 dB	0.600263% × € 161,100 × 0.02 × 110%	21.30
65–80 dB	1.68971% × € 161,100 × 0.02 × 110%	59.90
Air traffic		
54–65 dB	1.34962% × € 161,100 × 0.02	43.50
65–80 dB	2.30832% × € 161,100 × 0.02	74.40

6.12 Details of the Marginal Cost Calculations

The common procedure for simplifying the calculations of unit costs is to extract the main features of curvilinear exposure—effect relationships using piecewise linear interpolation. The slopes of each of the line segments are constant and can be used directly to indicate the change in the proportion of people affected by a 1 unit change along the x-axis.

6.12.1 Highly Annoyed

The Miedema curves (see Fig. 6.2) were approximated with piecewise linear segments with slopes of 0.942717, 0.600263, and 1.34962%, respectively, between 52 dB (rail 53 dB) and 65 dB. Above 65 dB the slopes are 2.12606, 1.68971 and 2.30832% for road, rail and aircraft noise, respectively. By multiplying each of the slopes with cost in DALY and the increase in influence area we obtain the noise impact cost of a 1 dB increase in noise exposure for the different noise sources taking the increase in influence area into account (see Table 6.7).

The additional fraction of people becoming highly annoyed by a 1 dB noise increase is lower when the initial noise level is under 65 dB than above.

6.12.2 Highly Sleep Disturbed

The exposure effect relationships or not too far from linear between 50 and 70 dB and the marginal changes below 50 dB are small (see Fig. 6.3). The slopes are approximately 0.5784% for road traffic noise, 1.5068% for train noise, and 1.1436% for aircraft noise.

Table 6.8 Marginal cost of sleep disturbances per dB (L_{night})

L_{night}		€2019
Road traffic > 50 dB	0.5784% × € 161,100 × 0.06 × 110%	71.70
Rail traffic > 50 dB	1.5068% × € 161,100 × 0.06 × 110%	186.90
Air traffic > 50 dB	1.1436% × € 161,100 × 0.06	141.90

When we use the disability weight of 0.07 and consider the increase in the influence area of a noise increase, we get the marginal costs (see Table 6.8) of an increase in exposure of 1 dB used in both Table 6.2 (L_{night}) and Table 6.3 (L_{den}).

6.12.3 Ischemic Cardiovascular Disease (Road Traffic Only)

The increase in RR pr increase in exposure of 1 dB is nearly linear with a slope of 0.00822225 (see Fig. 6.4).

Based on national health statistics the population risk for ischemic heart disease was 0.0007568 in 2016. This number includes both incidences in areas affected by different levels of noise, and quiet areas.

From the known fraction of the population that resides in each 5 dB noise exposure interval, it is possible to calculate backwards to find the incidence rates without noise and find the weighted average increase in population risk due to noise.

We found the fraction of ischemic heart diseases attributable to noise is 3.9% and the baseline risk of dying of ischemic heart disease without road traffic noise is consequently 96.1% * 0.0007568 = 0.00727258%. By multiplying with the number of DALYS lost per incidence 11.376 we find the marginal cost of an increase of 1 dB to be € 12.10 (see Table 6.9).

Table 6.9 Marginal costs per person per year due to myocardial heart disease from an increase in noise exposure (L_{den}) of 1 dB

Road traffic	Equation	€2019
L_{den} > 53 dB	0.00822225 × 0.0727258% × € 161,100 × 11.376 × 110%	1210

6.12.4 Moderate Annoyance

We use simple linear approximations of the exposure effect relationships (see Fig. 6.5).

The slopes are 15% per dB over 50 dB for road traffic and 1.50% per dB for rail traffic. The slopes for aircraft are $+0.00976928$ below 65 dB and -0.00880008 above 65 dB. The cost estimates then follow from the equations in Table 6.10.

Table 6.10 Marginal cost increase of becoming moderately annoyed (Calculation details)

Source	Equations	€2019
Road traffic noise $L_{den} > 50$ dB	$1.15\% \times € 397 \times 110\%$	5.00
Rail traffic noise $L_{den} > 53$ dB	$1.50\% \times € 397 \times 110\%$	6.60
Aircraft noise L_{den} 50–65 dB	$0.00976928 \times € 397 \times 110\%$	3.90
Aircraft noise L_{den} 65+ dB	$0.00976928 \times € 397 \times 110\%$	−3.50

References

Arsenio E, Bristow AL, Wardman M (2006) Stated choice valuations of traffic related noise. Transp Res Part d: Transp Environ 11(1):15–31

Basner M, McGuire S (2018) WHO environmental noise guidelines for the European region: a systematic review on environmental noise and effects on sleep. Int J Env Res Public Health 15(3):45

Bendtsen H, Fryd J, Popp C, Eggers S, Dilas J, Tønnesen A et al (2015) Guidance book on the integration of noise in road planning. Danish Road Directorate, Institute of Transport Economics, Lärmkontor, Copenhagen. Contract No.: ON-AIR Deliverable 4.1

Engelien E, Steinnes M (2021) More people exposed to environmental noise. Statistics Norway, Kongsvinger. https://www.ssb.no/natur-og-miljo/artikler-og-publikasjoner/flere-utsatt-for-stoy-ved-boligen

Klæboe R, Veisten K, Amundsen AH, Akhtar J (2011) Selecting road-noise abatement measures: economic analysis of different policy objectives. Open Transportation Journal. 4:87–92

Miedema HME, Oudshoorn CGM (2001) Annoyance from transportation noise: relationships with exposure metrics DNL and DENL and their confidence intervals. Environ Health Perspect 109(4):409–416

Mishan EJ (1988) Cost-benefit analysis: an informal introduction, 4th edn. Unwin-Hyman, London, UK

Nelson JP (2008) Hedonic property value studies of transportation noise: aircraft and road traffic. In: Baranzini A, Ramirez J, Schaerer C, Thalmann P (eds) Hedonic methods in housing markets: pricing environmental amenities and segregation. Springer New York, New York, NY, pp 57–82

Nilsson M, Bengtson J, Klæboe R (eds) (2014) Environmental methods for transport noise reduction. Taylor and Francis, CRC Press

Öhrström E, Skånberg A, Svensson H, Gidlöf-Gunnarsson A (2006) Effects of road traffic noise and the benefit of access to quietness. J Sound Vibrat 295(1–2):40–59

Øverland S, Knudsen AK, Vollset SE, Kinge JM, Skirbekk V, Tollånes MC (2016) Sykdomsbyrden i Norge 2016. Resultater fra global burden of diseases, injuries, and risk factors study 2016 (GBD 2016). FHI2018, Oslo

Pigou AC (1924) The economics of welfare, 2nd edn. Macmillan, London

Rødseth KL, Wangsness PB, Veisten K, Høye AK, Elvik R, Klæboe R et al (2019) The external costs of transport—Marginal damage cost estimates for passenger and freight transport in Norway. Institute of Transport Economics 2019, Oslo Contract No.: 1704/2019

Schultz TJ (1978) Synthesis of social surveys on noise annoyance. J Acoust Soc Am 64(2):377–405

Thune-Larsen H, Veisten K, Rødseth KL, Klæboe R (2014) External marginal costs of road traffic with revised accidents costs. Institute of Transport Economics, Oslo

Van Kamp I, Klæboe R, Brown AL, Lercher P (2015) Soundscapes, human restoration, and quality of life. Soundscape and the built environment. CRC Press, pp 43–68

Van Kempen E, Casas M, Pershagen G, Foraster M (2018) WHO environmental noise guidelines for the European region: a systematic review on environmental noise and cardiovascular and metabolic effects: a summary. Int J Env Res Public Health. 15(2):379

Chapter 7
Planning and Design Responses to Urban Sound—Learning from and Listening to Cities and Turning Knowledge into Sound-Aware Practice

Trond Maag and Arnthrudur Gisladottir

7.1 All Eyes on the City

People, buildings, open spaces and the urban morphology and infrastructure all contribute to the sound environment in cities. The crucial role of sound in improving the quality of life in urban areas is often underestimated, and efficiently integrating sound-related knowledge into urban planning and design practices requires collaboration among practitioners from different fields and the development of specific approaches and working formats. We will begin by exploring the interplay between sound and the city's built, natural and social spaces to approach the many facets of this topic.

7.1.1 Sounds Inherent in Urbanisation

The United Nations expects two out of three people to live in cities by 2050 (United Nations 2019). For many, cities hold the promise of jobs and housing, better access to health care and education, and social and cultural engagement, as discussed in Chap. 1. Cities also offer opportunities for achieving environmental, sustainability and health goals.

T. Maag (✉)
Federal Office for the Environment, Bern, Switzerland
e-mail: trond.maag@bafu.admin.ch

A. Gisladottir
NIRAS A/S, Aarhus, Denmark
e-mail: argi@niras.dk

© The Author(s) 2025
I. van Kamp and F. Woudenberg (eds.), *A Sound Approach to Noise and Health*,
Springer-AAS Acoustics Series, https://doi.org/10.1007/978-981-97-6121-0_7

The importance of well-planned urban development in providing suitable housing, public spaces, and efficient transportation infrastructure highlights the fundamental need for approaches to urbanism that will make tomorrow's cities more environmentally friendly, healthier and socially more inclusive. Those approaches also include ways to address environmental issues such as biodiversity loss, noise exposure and annoyance, air and water pollution, energy and land overuse, and the climate crisis. These changes are closely tied to the sound environment, as sound is inherent to space. The mechanisms of urbanisation, whether in inner cities or less populated regions, shape the environment acoustically by affecting the built and natural morphology and influencing the everyday behaviour of the people residing there. Urbanism and environmental design thus can make a massive difference to the urban sound experience.

7.1.2 Sound in Everyday Urban Life

The relationship between urbanism and environmental sounds is closely intertwined. Continuous urban growth brings the city and how people shape it acoustically through spatial organisation and urban design to all the places where people live, travel, work and move. Urbanisation concentrates services and infrastructure in urban cores, including central business districts and old towns. Often, a public transportation hub, such as a railway station, anchors the urban development around arts and cultural infrastructure, shopping and educational facilities. The proximity to homes, businesses, universities, schools and workplaces makes it much easier to reach people and amenities on foot or by a short bus ride. Living near city centres is attractive to many people, and already built-up areas may be changed for this purpose, for example, when former industrial areas are converted into housing and working developments. Space-intensive businesses, including warehouses and logistics providers, also prefer locations outside the denser neighbourhoods due to better freight and logistics connections along principal transportation axes. Land outside the urban cores tends to be more affordable, and car use as the primary transportation mode facilitates sprawling developments in these areas. As long as people create new land and infrastructure for agriculture, tourism and other (urban) purposes, urbanisation will also affect rural areas and landscape values. Sometimes, urban sprawl seems never-ending and irreversible (Maag 2013).

A few key factors shape the acoustic realities of compact and dispersed cities described above. In densely built-up areas, sound easily reflects between tall buildings, resonating and amplifying across public squares, streets, parks, and courtyards. The shape, size, scale, and materials of buildings, along with the outdoor design and layout, collectively influence the sound in cities. Construction sites, road traffic, and people engaging in street life and socio-cultural activities can add sounds to the urban cacophony, further augmented by the sounds from aircraft, ferries, boat traffic, air conditioning units, industrial operations, and commercial places.

Sound can travel longer distances in areas characterised by dispersed buildings and more open spaces, affecting a broader area than in denser urban typologies. A city's development stages, which over time determine density, traffic patterns, the ratio of unbuilt to built-up areas, and settlement forms, significantly influence noise propagation (Margaritis and Kang 2016, 2017). Wind and temperature gradients, driven by climatic effects and the terrain's morphology, also play a significant part in the acoustic interplay of open urban spaces. As the terrain becomes more open, often extensively used for purposes such as agriculture, energy production, and tourism, such sounds of peri-urban and rural areas can influence its development options.

Sound quality influences whether people prefer to live in a particular area and are more likely to pay a higher property price or stay longer in a specific location. Environmental influences, including sound, can contribute to the segregation of neighbourhoods, impacting people's health, education and access to public services (Leiper and Hood 2022). People with higher incomes have better resources to avoid negative influences (Diekmann et al. 2023).

7.2 Three Perspectives of Sound on Cities

This section discusses three ways of approaching sound-related urbanism to address the context outlined earlier. Two well-established practices have emerged in urbanism. The first, widely accepted approach is to minimise and manage unwanted sounds and harmful sound levels to avoid potential health risks. The second approach aims to shape the sound environment by enhancing positively perceived sounds, promoting well-being and creating recreational environments. In contrast, a third, less conventional approach takes a perspective on planning and design practices. It seeks to efficiently exploit synergies, recognising that most sound-related actions are woven into existing planning and design processes across disciplines and protocols (Brown 2021).

7.2.1 Improving Public Health Through Reducing and Modifying Noise Levels

Environmental sounds affect mental and physical health, and severe health risks, such as sleep disturbance and cardiovascular diseases explored in Chap. 4, have been associated with long-term exposure (Basner et al. 2014; World Health Organization 2018, 2011). This particularly negative connotation of sound, commonly termed noise, makes the maxim "the less noise, the better public health " a simple but credible principle for urban spaces where potentially harmful sounds, in this context labelled as noise, are ubiquitous (Fink 2019). The enormous efforts needed to control noise pose a significant challenge, especially if it is overlooked in the planning process. In recent decades, spatial planning practice and legislation have developed three interrelated strategies to reduce the health risks associated with noise, as this section will show. These strategies have strongly shaped what is commonly known as environmental noise management, discussed in Chap. 8, and encompass a comprehensive range of interventions (Brown and Van Kamp 2017) qualified and assessed through environmental noise mapping.

7.2.1.1 Land Use Planning and Master Planning

Planners and developers use several ways of responding to noise, especially to comply with regulations and specify locational issues for buildings and activities sensitive to noise, such as residential and recreational areas. In land use planning, the general design intention is to keep noisy activities at a distance from sensitive ones. Master planning intends to achieve a balance between different interests of a specific area, such as housing, work, travel and recreation, through the layout of the buildings, arranging the floor plans and positioning the outdoor spaces. However, challenges arise when planning areas in airport regions or densely built-up areas where smaller apartments face busy streets. When noise-sensitive uses cannot be reduced to certain levels, the noise department of the municipal authority may agree to grant exemptions if adequate regulations are in place through the permitting process. For example, better acoustic insulation can prevent exposure to traffic noise inside the building. Combined with specific requirements for outdoor spaces, such as a quiet inner courtyard, it can compensate for the noise on street facades. These measures effectively ensure people's health, especially in consolidation and inner city areas, but do not reduce noise itself.

7.2.1.2 Planning and Design of Transportation Infrastructure

City planners and urban designers are not directly involved in designing car engines, trains and aircraft. Instead, they design the infrastructure where these machines are used and frame how people use it, ultimately impacting the resultant sound.

Planners can address road noise by using quieter asphalt pavements for roads. Slowing down traffic, altering traffic patterns and restricting heavy vehicles and night-time traffic are also highly effective in reducing road noise, especially when combined (Estévez Mauriz et al. 2016). Well-maintained wheels and rail grinding are often sufficient to address railway noise, especially when combined with proper track construction, quiet locomotives, and trains fitted with disc and composite brakes. Various factors related to aviation, such as ground handling procedures, operating hours at airports, landing fees and flight paths for landing and takeoff, are critical in addressing and mapping aircraft noise.

Using screens and walls as part of the infrastructure is a promising way to lower noise, with a more extensive design closer to the source providing a better shielding effect. However, in inner city areas, the massive construction of noise barriers can severely interfere with public spaces and discourage pedestrians and cyclists from using them. Although simple walls and barriers are common in private gardens, they are often too small to significantly reduce noise, as the lower frequency spectrum with wavelengths up to 17 m is challenging to control with small buildings and lightweight constructions. However, adequately designed low walls with specific acoustic values and a visual appeal can be an option, as seen in waist-high walls in Helsingborg made of differently designed brick surfaces with an absorbent core (AFRY, Efterklang 2020).

Modifying potentially noisy infrastructure involves promoting walking and cycling, expanding public transportation, densifying networks and redesigning streets. Municipalities can contribute by investing in quieter machinery, such as garden tools and maintenance equipment, or upgrading public transportation fleets through public tendering. As with saving water and energy, they can encourage people to drive less noisy vehicles. For example, they can offer incentives to residents who use electric cars, such as free parking, adequate charging facilities and free use of bus lanes. They can also draw attention to the benefits of low-noise tyres. Additionally, national authorities can support these efforts on a larger scale by funding quieter technologies and transportation through incentive schemes.

Reducing noise often requires massive efforts, and more than one measure may be necessary to reduce it to acceptable levels for those living near the source. However, combining different measures can have a positive effect and benefit people living further away. This approach is important since reducing traffic noise by around 3 dB(A) is needed for people to perceive a noticeable difference, which is equivalent to halving the traffic volume on roads with mixed traffic or reducing the speed from 50 to 30 km/h. While the lack of quick fixes is evident and may discourage local authorities from initiating projects, each contributing a small effect, a combination of measures is crucial for achieving lasting noise reductions (Brown 2021).

7.2.1.3 Urban Morphology and Architectural Elements

The frequency range of around 20 Hz to 20 kHz relevant for most noise sources corresponds to wavelengths between 17 m and 17 mm, meaning sound waves interact directly with the typical dimensions found in cities. These interactions include acoustic phenomena such as reflection from walls and diffraction at edges (Hornikx 2016). City planners and urban designers have a unique chance to control and manage noise by considering various aspects of the urban morphology, including the location and orientation of buildings, the (horizontal) layout and (vertical) profile of streets along buildings, as well as surfaces and materials of ground coverings, buildings facades, and the terrain and vegetation of urban spaces (Gisladottir et al. 2018).

Planners and designers should initiate specific actions to promote this approach. Landscape architects, for example, can phase out the asphalt in non-traffic areas and replace it with planting beds to improve sound absorption. Adding water features and street furniture can also help create a more lively urban space. Also, architects play a significant role in urban morphology by integrating vegetation into building facades

Fig. 7.1 The shape of facades and roofs, the materials used in streets and public spaces, the arrangement and height of buildings, and the measures implemented to control noise sources collectively shape the sound experience of a specific location in the city, influencing how people perceive and use that particular location. Photo credit: Trond Maag

and roofs. They can also structure building surfaces into smaller units and bend and position them to avoid unwanted reflections. This aligns with the design responses of architectural acoustics, where techniques like surface texturing and room structuring are employed through acoustic modelling to achieve sound qualities in indoor spaces such as concert halls and theatres. As later discussed, similar design and architectural approaches can significantly improve the perceived sound quality of public spaces in the city context (Fig. 7.1).

7.2.1.4 Environmental Noise Mapping and Monitoring

The health benefits of reducing noise exposure are evaluated through noise monitoring and mapping, as regulated in the EU *Environmental Noise Directive* 49/2002/ EC, *END* (European Parliament and the Council of the EU 2002). The relevant noise indices and limit values are discussed in Chaps. 2 and 8. Operating at a higher-level planning scale within the *Strategic Environmental Assessment (SEA)* process, noise mapping enables planners to test policies and scenarios for adjusting the movement of goods and people, for example, through redirecting transportation routes and redesigning infrastructure, as well as modelling public intercity transportation ticketing and road pricing (Brown 2021). It also helps in formulating urban densities and land use policies relevant to noise, including areas for tourism, wind farming, industrial facilities, and recreation. Noise is just one of several environmental factors within SEA and is rarely critical in introducing or modifying land use and planning policies. However, SEA can serve political discussions and facilitate public debates concerning the balance between spatial and noise-related interests, such as the case of urban consolidation. For example, a political motion (Swiss Parliament 2018) prompted a requirement for new legislation to adjust Swiss noise regulations, aiming to facilitate urban development in built-up areas (Swiss Federal Council 2022).

Noise mapping also plays a crucial role in the *Environmental Impact Assessment (EIA)* process, where noise is considered alongside soil, air, water and other environmental aspects, all examined on a project-specific basis. As a pivotal planning tool, noise mapping evaluates and optimises an individual project within a designated place and, subsequently, takes it through the planning approval stages. Noise mapping techniques, such as those used in SEA and EIA, model noise employing Geographic Information System-based mathematical and statistical data. Planners and developers may also use these data, tailored to a specific project's regulatory process, to meet a building's noise requirements, implement measures within a master plan to protect noise-sensitive receivers or facilitate upgrades, redesigns and retrofits for infrastructure that fails to meet noise standards.

The ways individuals respond to sounds within a given context play a pivotal role in overall quality of life. Planners and developers need to address this issue by controlling noise levels to contribute to long-term health benefits while facilitating local sound experiences. Planning and design practices should promote the sound qualities of places and facilitate acoustically positive experiences to achieve the latter. We will explore this aspect in the next section.

7.2.2 Promoting and Shaping Sound Environments that People Prefer Over Noisy Places

The sounds we hear in a particular place, combined with tactile, visual, climatic and other sensory experiences, play an essential role in shaping the perception of space and influencing people's experiences within it. They influence how a particular person responds to and interacts with the environment. Sound is also massively involved in how stressed or relaxed people feel in a particular location. City planners and urban designers can cultivate certain qualities around this inherent connection between place and sound. The soundscape defined under the ISO framework (International Organization for Standardization 2014) refers to any sound environment that is contextually valued, whether a busy and roaring city market or a serene natural setting. Approaches to considering and implementing soundscape knowledge into environmental design and urbanism trace back to the 1950s and 1960s, particularly through perceptual walks aimed at sensing urban spaces (Lynch 1960; Southworth 1969).

7.2.2.1 Sound as a Resource for Urbanism

Soundscape practice is scientifically integrated today through diverse methods such as laboratory experiments, narrative interviews and behavioural observations (Aletta et al. 2016), with distinct forms for evaluating and communicating results. They link contextual cues and social relationships (Axelsson et al. 2010; Axelsson 2015; Kang et al. 2016) positively associated with the sound environment rather than discomfort and disturbance (Brown and Muhar 2004). The perceived sound environment cannot be described in absolute terms, emphasising the need for consensus on measurement and evaluation procedures and best practices (Aletta et al. 2016; Brown et al. 2011).

Soundscape knowledge is gaining significance in urbanism, at least in Europe, where sound quality factors beyond just absolute metrics are generally accepted to play an essential role in enhancing well-being, health benefits and quality of life. Soundscape results are increasingly being incorporated into projects and plans, mainly through guidelines and standards, such as the non-mandatory ISO standard and the mandatory END noise action plan. Typical objectives are sound qualities that enhance the urban experience and compensate for places perceived as noisier. For example, planners and developers can promote sounds that people prefer to hear and create local experiences supported by factors of quietness and tranquility in the given location (Maag 2016). Virtual Reality equipment can be helpful during the design process (Echevarria Sanchez et al. 2017). For example, Yanaki and colleagues have developed an immersive soundscape sketchpad for prototyping soundscapes in urban public spaces (Yanaky et al. 2023). Furthermore, acoustic modelling and planning scenarios, as exemplified by Kropp and colleagues in the SONORUS project across different European cities, can be valuable in making informed decisions to improve the sound environment (Kropp et al. 2016). Knowledge of soundscape actions needs to be further translated into practical design methods and protocols, as noted by Cerwén (Cerwén et al. 2017).

7.2.2.2 Quiet Areas and Acoustically Coherent Public Spaces

Quiet areas are an example of a planning tool within the EU where soundscape knowledge comes into play. Cities and agglomerations must designate quiet areas regulated under the END as part of noise action planning. Principles and rules of thumb for areas experienced as pleasing and restorative have been introduced by the European Environment Agency for "quiet areas " (European Environment Agency 2016, 2014), the HOSANNA project for "quieter and greener cities " (Hosanna 2013), and the QUADMAP project for "quiet urban areas " (Quadmap 2015).

Noise action planning usually assesses the potential quiet areas' noise levels, purpose, minimum dimensions/size and location (Heinrichs et al. 2018). This approach can emphasize areas that provide opportunities for restorative and quiet experiences, such as parks and green spaces, forests, agricultural land, water bodies and brownfield sites. Planners prefer to designate quiet areas where (long-term) noise levels are already low. However, large areas assigned that way may thus be located far from densely built up areas, primarily if a minimum size criterion discourages smaller areas that could offer relief from everyday noise in the living and working environment (Senatsverwaltung für Umwelt, Verkehr und Klimaschutz 2020).

Open spaces of all sizes and purposes, including courtyards, gardens, multifunctional and pedestrian-friendly spaces and public squares, can be correlated with criteria for developing quiet areas. Geographic Information System-based techniques can assist planners and developers in identifying suitable locations for potential public quiet places based on factors such as size, accessibility and design. However, residents may have experiences and needs that are not initially recognised by planners but are critical in accurately aligning and designing quiet areas (Payne and Bruce 2019). Determining quiet areas thus always includes involving the public, which can be done through participatory methods online, such as (Radicchi 2018), or face-to-face in user surveys (Bonacker and Bachmeier 2018). By creating new open spaces and upgrading existing ones, increasing the degree of public accessibility, promoting opportunities for recreation and exercise and interconnecting such spaces for pedestrians and cyclists, planners can respond to people's recreational and leisure needs. Interwoven spaces can connect differentiated sound qualities of larger and smaller places, creating an acoustically coherent and highly accessible public space (Fig. 7.2).

Fig. 7.2 The redesign of the Hovinbekken river in Oslo, Norway, aimed to improve the area's drainage and stormwater efficiency. A river walk leads people through residential areas, creating connections to public transportation and nearby recreational spaces over a length of two kilometers. Photo credit: Trond Maag

7.2.2.3 Improving the Perceived Sound Environment

In cities, there are always unexpected, surprising sounds, such as people telling stories, children laughing, footsteps joyfully echoing from buildings or water moving erratically, that can capture the listeners' attention and bring a smile to their faces. Over time, each place develops its own acoustic identity (Fig. 7.3). This identity can even contribute to the city's collective memory, and its acoustic icons can also hold cultural values (Maffei et al. 2016; Bartalucci and Luzzi 2020). Socio-cultural factors and spatial context influence the perception of the sound environment and play a significant role in how people appreciate and cope with their environment (Van Kamp et al. 2016). Studies, such as (Fryd et al. 2016), show that noise annoyance decreases when people associate the noise sources with their neighbourhood. More anonymous noise sources, such as motorways, lead to higher annoyance for the same noise level. Another study (FAMOS 2022) indicates that city residents experience less noise when they know transportation infrastructure is safe and clear than when it is not and that the design and visual appearance considerably affect annoyance.

Especially in light of urbanisation, there is a need to design spaces that individuals and communities perceive as more peaceful and restorative. Decibels may be the most apparent modifier to achieve this. However, senses other than listening,

Fig. 7.3 The port of Hamburg, located in Germany, can be a busy and noisy place. However, technology plays a role in reducing some of the noise sources; for example, sensors are used to automatically slow down containers during handling, which reduces the noise associated with their placement (personal interview with Christian Popp (Maag 2013)). Implementing such measures contributes to the development of urban areas and the promotion of the acoustic identity of the port city, both of which are vital for Hamburg as a tourism destination. Photo credit: Trond Maag

particularly those related to vision and motion, have also been considered relevant to constructing a tranquil space (Pheasant et al. 2010). It is important to note that these sensory experiences, especially from green and water spaces, have benefits that cannot be entirely substituted by noise-reducing measures alone (Van Renterghem 2019; Vienneau et al. 2017). For example, Schäffer and colleagues show that people perceive roads and railways as up to 10 dB(A) less noisy the greener the neighbourhood (Schäffer et al. 2020). The (visual) analysis of green metrics, gathered from satellites, Geographic Information System data and street-level observations, is used in acoustic research to highlight the importance of green spaces and natural environments in cities (Leereveld and Margaritis 2023).

Green spaces also contain audible resources from wind, plants, insects, and birds, which improve spatial orientation and add variety to listening situations. Sounds that are more attractive and exciting to the ear can help to keep unpleasant and unwanted sounds in the background of perception. Moving water often has a positive effect; its acoustic resources provide an acoustic wealth many listeners appreciate and find enjoyable (Brown and Muhar 2004).

The visual appearance of an environment is a strong modifier of how people value the urban sound experience. It points out the supportive and unfavourable factors planners and developers are encouraged to consider. Some factors will help sharpen and differentiate the perceived sound environment. In contrast, others will interfere with a location's acoustic uniqueness and reduce the urban sound experience. Planners and developers can also evaluate design scenarios by balancing unfavourable and supportive factors. For example, a more resident-friendly immediate environment could offset the perception of traffic noise in residential buildings. The positive influence of certain design factors provides planners and developers with some leeway to compensate for noise (Hallin et al. 2006).

It is worth noting that a city's culture and history strongly influence the perception of and the design responses to sounds. A study (Amphoux et al. 1991) exemplifies this by examining the different sonic milieus of three Swiss language regions to provide insights for describing the sound quality of European public spaces. It reminds us that understanding sound is inherently tied to the disciplinary perspective from which it is studied and varies across cultures and contexts (Krogh Groth and Mansell 2021). For example, Aletta and colleagues show differences in how people in Europe and China assess and value city soundscapes (Aletta et al. 2023).

7.2.3   Developing Places Through Synergies with Urban Sound

Given the complexity of sound, it is essential to communicate and transfer knowledge in formats and languages that are accessible to practitioners in relevant fields (Steele 2018; Taylor and Hurley 2016; Gisladottir 2021). Practitioners, especially in urbanism, architecture, and environmental design, typically consider environmental noise objectives and qualitative and contextual aspects separately or overlook both. The movement towards more integral design approaches in these disciplines thus presents an opportunity to integrate specific knowledge about environmental sounds into the design process, linking it closely to projects and protocols concerning other objectives, such as improving air quality and reducing heat island effects (Pont et al. 2023). This way, planners and developers achieve multi-criteria goals efficiently. It is also essential to take into account the city's culture and history, as they strongly influence the perception of sound and the design responses to it. Overall, the holistic perspective on sound as an expression of urbanity makes the urban sound a multi-layered quality that planners and developers can emphasise through non-sound-specific practices.

Even subtle design choices can considerably influence the acoustics of a given location (Taghipour et al. 2019). When modelling architectural outdoor and public spaces, environmental noise mapping techniques often fall short, failing to capture the details typical to the architectural scale and its complex acoustic phenomena, such as reflections between surfaces and diffraction along the building edges. Hornikx suggests using complementary methods that better account for architectural (micro)scales when designing urban areas (Hornikx 2016). Acoustic models with a higher computational resolution, i.e. full-wave numerical methods, can handle dimensions as small as a few millimetres. Ideally, they can also account for moving noise sources, which is significant in scenarios like designing roof shapes along roads where noise sources are in motion (Van Renterghem and Botteldooren 2010). Research from Eggenschwiler and colleagues shows that the perceived sound quality can change due to different materials and building constellations in urban spaces with moving vehicles (Eggenschwiler et al. 2022). The quality rating is attributed to both sound pressure levels and sound character variations. Van Renterghem and colleagues show that solid, acoustically rigid surface materials can elevate traffic noise levels compared to their softer counterparts (Van Renterghem et al. 2013). Similarly, Echevarria Sanchez and colleagues shed light on how noise levels in urban street canyons are tied to the geometric profiles (Echevarria Sanchez et al. 2016). Beyond considerations specific to the built morphology, descriptors for landscape patterns (Han et al. 2018) and ecoacoustic metrics (Haselhoff et al. 2022) offer additional avenues for research and provide insights for finding design responses that align with the challenges posed by urban spaces.

The public urban sound realm emerges both physically and socially. It arises physically, from sound generation and propagation within the urban fabric, and socially through people's behavioural responses to the city sound. A growing understanding of this dual nature underlines the importance for planners and developers to consider proactively people's engagement with their environment. This consideration encompasses different approaches, such as incorporating artworks and cultural infrastructure or unlocking acoustic co-benefits through social programmes. These approaches involve working with local communities and cooperating closely with local authorities and developers to address questions surrounding who and how people use particular spaces, what people expect from a place and how they can actively participate in shaping these areas. Projects adopting these can provide surprising insights into environmental sound and noise issues, avoiding overly academic and technical language. For example, the City of Utrecht in the Netherlands prioritises health in its municipal policy, building a solid case for a healthy (sound) environment. This approach allows projects to be linked to environmental sounds, as the Healthy Urban Living for Everyone programme exemplifies (Gemeente Utrecht, Healthy urban living for everyone. URL accessed 15 April 2023: https://healthyurbanliving.utrecht.nl/).

Minimising noise, promoting positively connoted sounds and recognising opportunities for acoustic synergies. These approaches discussed above deserve wider attention, as they must be adequately addressed by more than one discipline. The following section explores ways cities can better benefit from these approaches and integrate them more seamlessly into planning and design practices.

7.3 Planning Contexts

To effectively achieve the multiple objectives of cities, there is a convincing argument for placing greater importance on sound-related considerations in urbanism and environmental design. This section emphasises the strategies to create more compact cities, the crucial role of public spaces in achieving this, and the opportunities presented by aligning planning and design practices with broader sustainability objectives, particularly those outlined in the UN's Sustainable Development Goals. The key idea is that efficiently integrating environmental sounds into these planning contexts can enhance coordination with existing efforts, ultimately contributing to the urban sound experience.

7.3.1 Consolidation Strategies of Cities

The European Environment Agency monitoring programme estimates that at least 20% of the population is affected by harmful traffic noise (European Environment Agency 2019). A Swiss study indicates that over 90% of people affected by environmental noise live in urban areas, with one in eight affected at night and one in seven during the day (Catillaz and Fischer 2018). While cars are the most common noise source in urban areas, aircraft, railways, and other infrastructure can have concentrated impacts. Although there are examples of progress in reducing environmental noise, the number of residents affected by noise is expected to remain high (European Environment Agency 2022).

In many cities, the current urban development paradigm intends to limit new infrastructure and buildings to certain areas. This inward-oriented planning aims to protect the land from inefficient use and urban sprawl, reduce traffic and use fewer resources for housing, work and leisure. Urban consolidation, which involves more people living in densely populated cities, leads to more noise from transportation infrastructure, leisure activities, tourism, and cultural and sporting events. The noise department of the municipal/local authority may grant exemptions for permitting buildings and infrastructure, even if it involves a certain derogation from health objectives. Planning and building authorities are unlikely to refuse permission solely based on environmental noise regulations. Planners and developers can physically reduce the noise by improving the insulation, enclosing balconies, and taking other measures on the building envelope. Residents inside apartments appear to be protected from external noise, but such measures create a disconnection from the outside as residents cannot hear it. The external noise remains in the area and can limit the possibilities for further development. The high pressure on local authorities to grant permissions for planning and construction, especially in areas with high demand for housing, emphasises the need for a more proactive urbanism response to sound.

7.3.2 Developing the City Through Public Spaces

By concentrating people, buildings and activities in a limited space, cities compete with other cities for professionals and students, residents and tourists, and businesses. Cities thus focus on these demands by providing, for example, spectacular buildings, health infrastructures, fast mobility options, and attractive cultural and sporting events. At the same time, cities must meet people's needs for family life, leisure and recreation. Efforts in this regard include pedestrian and cycling infrastructure, shared spaces and residential streets, and resident-led projects for human-scale spaces and sub-cultural activities. Public spaces, from the quietest to the noisiest, including squares, streets and parks, and their design are prerequisites for creating attractive cities.

Meeting diverse demands in limited spaces presents an acoustic challenge to urbanism accompanied by addressing environmental objectives related to city climate, urban ecology and biodiversity, as well as social objectives. Integrating sound quality into design practices that consider the various disciplines intersecting in the public realm can facilitate the acceptance of consolidation strategies in dense urban areas. Suter and colleagues found in the Swiss urban development context that the more the design efforts contribute to making compact spaces quieter, the better the consolidation strategies are accepted (Suter et al. 2014). Similarly, drawing people's attention to landscape values through focusing on sensory experiences such as being away from home, fascination with a particular place, the extent of feeling connected to it, and compatibility with one's activities (Kaplan 1995) may encourage planners and developers to consider the urban sound experience more holistically. Furthermore, social functions and recreational experiences can moderate the perception of noisy environments (Gidlöf-Gunnarsson and Öhrström 2007). People value specific sounds, such as sounds from moving water and vegetation, and access to forests and parks (Irvine et al. 2009). These elements are associated with green spaces and the landscape, serving as strong visual icons of an acoustically attractive city. City dwellers value these icons, and urban designers are encouraged to enhance their impact by creating emotionally more resonant public spaces, enabling people to explore them on foot or by bicycle.

7.3.3 Unlocking Urban Synergies Through Sustainability Approaches

The built, natural and social environments undergo profound structural changes as cities grow. The United Nations General Assembly (United Nations General Assembly 2015) has established the Sustainability Development Goals (SDGs) as a guide to navigate these changes and address global challenges relevant to everyday urban life, such as climate crisis, biodiversity loss, socio-economic segregation and social deprivation. The SDGs encompass 17 high-level objectives with implications for research, legislation and building codes.

Looking more closely at the SDGs, we see that environmental noise only plays a role in the European edition of SDGs 3 and 11 (European Commission 2018). Unsurprisingly, SDG 3, which stands for "good health and well-being ", mentions noise as an essential health factor. The SDGs also link environmental sounds to more qualitative factors of sustainable urban development. Primarily, SDG 11 on "sustainable cities and communities "provides an opportunity to go beyond noise reduction strategies by paying attention to the overall urban sound experience. Environmental sounds can also matter in several other SDGs. For example, promoting "clean water and sanitation" in SDG 6 involves improving the water quality of lakes, streams, shorelines and rivers. Biophilic design can also help create recreational spaces that people find acoustically pleasing. Sustainable infrastructure, promoted by SDG 9, and land preservation, promoted by SDG 15, also have profound but usually little regarded synergies with environmental sounds.

Due to the many overlaps between the SDGs and sound, working with SDGs could pave the way for urbanism to identify synergies from which sound can benefit. A holistic understanding of the dynamic city driven by socio-economic and environmental mechanisms can facilitate such synergies. However, there is also a risk that such goals will not be achieved, as noise has yet to be integrated into the formulation and setting of sustainability goals, apart from the European context mentioned above (King 2022).

7.4 Sound-Aware Urbanism

Urbanism and environmental design need to move towards maximising synergies to enhance co-benefits for sound quality. This requires understanding the planning and design process and the individuals involved in making decisions related to sound. The first part of this section sheds light on formal procedures and challenges of incorporating sound into urban planning and design. Sound could take a critical, site-specific share by making planning more aware of everyday experience and participation, as explored in the second part.

7.4.1 Potential and Challenges of Urban Planning and Design Practices

Urban planning and design take place at different spatial scales and levels of decision-making involving a variety of stakeholders (Gisladottir et al. 2018). The process is embedded in hierarchical planning laws and building codes set at municipal/local and national levels. It seems helpful to use simple examples to give an insight into a process that often does not proceed linearly but over a long period with setbacks and delays.

At a higher level planning scale, planners and developers identify the urban area's primary objectives and purposes in a general and conceptual way using appropriate planning tools such as the spatial concept and the zoning plan. Already at this city scale, these tools may contain critical design intentions that affect the sound environment and determine which sounds and how people respond to them are more likely to occur. Initial steps to influence the sound environment positively at an early planning stage include:

- identifying and designating the location and purpose of noise-sensitive developments, such as residential areas and recreational spaces, and noise-generating ones, such as transportation and industrial infrastructure and commercial developments;
- achieving spatial densities (inhabitants per m^2, m^2 of open space per inhabitant) and indices such as compactness ratio and degree of land/building use, which influence how people get involved in activities in public spaces;
- identifying and designating urban spaces not to be used for development, which can, to some extent, preserve and maintain existing (sound) qualities;
- establishing large city parks, which serve both as anchors for development and long-term recreational infrastructure, especially when located in proximity to public transportation hubs;
- setting the pedestrian areas and cycling networks, which facilitate people's access to quieter neighbourhoods, their interaction with urban infrastructure and their daily routines using the city.

This macroscale planning does not consider the urban fabric and morphological factors in detail. Decisions at this scale aim to direct feasible developments with environmental and socio-economic improvements. For example, they may be concerned with the location of a hospital or a housing development for a certain number of people. The emphasis is on the "what and where " rather than the "how of design ".

However, the decisions made at this scale have acoustic consequences at minor planning scales because they play a role in framing the spatial and social characteristics of the context.

Master plan decisions typically involve an area's urban fabric and morphology, which may cover a prominent part of the city with multiple landowners. At this neighbourhood scale, planners and developers typically develop the buildings and the public spaces that affect how sound propagates into and through the urban fabric. Design decisions based on environmental noise mapping are often necessary, especially for noise-sensitive developments, in order to obtain permission for master planning and, subsequently, for the individual buildings. Although the surfaces of the buildings and the terrain will influence the quality of the sound environment, design teams do not necessarily specify the details at this scale. Finer details, such as the texture and structure of the surfaces, building materials, landscaping, ground materials, and vegetation, can all be left open. Large buildings and projects often involve several design teams working on specific sites to be developed in parallel according to a master plan. Final design decisions for facades, materials and outdoor spaces are often deferred to a later stage in the building design brief.

It is easy to imagine the city planners, urban designers and many other professionals involved over a long period, from sketching out initial ideas to constructing a site. However, sound-related qualities often depend heavily on the initial stages of the process, when decisions about the buildings' shape, height, position and size are made. Architects have little flexibility to modify sound-related design intentions beyond what is specified in the master plan and building design brief. City planners and urban designers indirectly influence the sound environment through their plans and programmes for objectives unrelated to sound, emphasising the importance of linking considerations of sound to other urban design objectives (Maag et al. 2023). Suppose the developer does not think about environmental sounds or does not explicitly ask for certain qualities. In that case, the chance is high that they will not be included in the master plan and the building design brief and will be overlooked altogether.

The early decisions made by planners and developers can easily dictate the sound environment of a particular area, which cannot be easily changed at later development stages. For example, changing the materials for a building's facade cladding may be simple to implement through a city climate programme. However, it will have much less impact on the sound environment than changing the building's form and shape. Another challenge is that decisions can change at any point in the process, for example, triggered by expert analysis, court cases or the involvement of new developers and landowners. Given the complexity of the planning and design process and the nature of the urban sound realm, it can be challenging to prioritise sound-related design intentions (Maag et al. 2019). Planners and developers often have no choice but to adhere to regulations and follow environmental noise mapping and mitigation practices. This leaves limited room for alternative approaches, such as listening experiences from walks, questionnaires, soundscape studies and considerations from acoustic modelling, design scenarios and Virtual Reality simulation, which could influence the development process towards better sound qualities at an early stage.

7.4.2   Co-Caring and Co-Sustaining the City by Citizens

In shifting the perspective from the planning and design process outlined above to that of everyday experience, people become connected to the formal process and actively contribute to shaping its acoustic consequences. As they feel and take responsibility for spaces, people interact acoustically with the city through their social and cultural practices, alone and shared with others (Ouzounian 2014; LaBelle 2019). People's presence, intertwined with daily routines and interactions in urban spaces, influences the environment's identity in social, cultural and acoustic terms. Recognising and understanding these mechanisms of co-shaping urban spaces helps to become more familiar with people's expectations for these places and how they perceive sound within a particular location. Getting to the bottom of how people interact with their environment is essential. How they move around, the routes and spaces they use, the pauses they make when walking and cycling—these social cues provide insights into whether people experience certain parts of the city as more fragmented or coherent acoustically. This understanding, in turn, facilitates specifying sound-related objectives and framing the formal planning process in a way that better resonates with the needs and expectations of the users.

Involvement of a place's users is crucial in order to incorporate sound in planning, build acceptance for a project and enhance confidence in planners, developers, and the municipality. It can help avoid conflicts later in the development process (Bonacker 2018). Involving the public in a participatory manner and aligning design intentions with their needs can make a location more pleasant. Consequently, planners and developers cannot simply follow the protocols of the formal planning process. They must use appropriate tools to effectively communicate with communities and local authorities, addressing contextual and site-specific issues. Walks, for example, allow for simple participation that is accessible to everyone, without using specialist language or equipment and questionnaires that only some are familiar with (Radicchi et al. 2017). They also aid in establishing simple reference points for encountering the sound environment from a personal perspective (Maag et al. 2019). Such work with the public, short-term or spread over several years, can help to:

- gain impulses from residents that can co-frame the formal planning and design process;
- synchronise objectives and share approaches to urban sound and noise between the disciplines involved in the formal process;
- promote a degree of self-responsibility for the sound environment in order to develop more sound-aware design responses;
- cultivate a sense of self-empowerment for the sound environment among residents and visitors;

- create a public sound realm people can identify with and thus relate to and cope with more effectively.

Three examples below can demonstrate the implementation of these principles. They illustrate that residents and experts can cooperate in many ways, such as participation, co-creation, public engagement, collaboration with artists and public artworks. It is up to the planners and developers to take advantage of the opportunities offered by such approaches, where people's everyday experiences and individual contributions are considered integral to the formal planning and design process.

A participatory process accompanied the redesign of the *Kirsebærlunden* playground in Oslo, Norway (Bymiljøetaten Oslo kommune 2022, Kirsebærlunden lekepark på Tøyen. URL accessed 16 April 2023: https://www.youtube.com/watch?v=4TFTZ9muVbM). The playground is situated between residential buildings and a hill, providing a quiet space close to a noisy street and a busy metro station. In response to public initiatives, the municipality installed three slides suitable for all ages and planted fifty trees, creating a unique environment with the voices of playing children and visitors (Fig. 7.4). Inspired by the residents, the design was guided by Oslo's programme for car-free city life (Oslo kommune 2023, Bilfritt byliv—byutvikling. URL accessed 16 July 2023: https://www.oslo.kommune.no/byutvikling/bilfritt-byliv/). Other programmes and plans in Oslo have also resulted in attractive sound environments, such as redesigning the *Hovinbekken* river mentioned above as part of Oslo's rainwater strategy (Bymiljøetaten Oslo kommune 2020, Nå sildrer Hovinbekken igjen, etter 120 år i rør! URL accessed 16 April 2023: https://www.youtube.com/watch?v=TyUEdf83wxE).

Working with artists in public spaces can invite people to enjoy exploring these spaces, for example, through permanent sound installations or temporary public interventions. Artworks and cultural infrastructure can create poetic views of spaces, encouraging people to push the boundaries of their usual imagination and perception of places. For example, *The Sonic Gathering Place* in Melbourne, Australia, created by artist Jordan Lacey in 2022, is a circular seating area designed with plantings and speakers playing back recordings (Lacey 2022). Its design combines biophilic principles and natural field recordings. This permanent sound art installation provides a unique experience for people in an otherwise busy urban fabric, offering a rare respite through access to natural sound immersion. It also serves as an urban laboratory platform for public discussions, studies, and exhibitions on sound and urbanism (Fig. 7.5).

Artworks and cultural infrastructure also help people reinvent places by actively taking acoustic control of specific locations through sound. Artist Sven Anderson demonstrated this through *Continuous Drift* in Dublin, Ireland, a public sound installation developed in 2015 (Fig. 7.6). This artwork allows people to use their mobile phones to trigger different sound recordings contributed by over 30 invited artists, played back through loudspeakers integrated into the architecture of a central public square (Anderson 2018). A few years later, he developed a temporary sound garden in Struer, Denmark, in collaboration with Trond Maag and Andres Bosshard (Maag 2021). In this sound garden, school classes learned about the relationship between

Fig. 7.4 Kirsebærlunden playground was redesigned as part of Oslo's car-free city programme. Residents' design ideas were considered by actively involving the public in the design process. Photo credit: Trond Maag

sound and architectural space using boomboxes loaded with analogue cassette tapes featuring recordings of local sound environments. By working together to choose the cassettes and position the boomboxes, the students became involved in the acoustic design of the sound garden. Alongside this engagement with acoustic design, the sound garden also served the students as a platform to explore Struer's heritage and current identity as a city of sound (Struer Lydens By 2023, City of Sound. URL accessed 16 July 2023: https://cityofsound.dk/).

7.5 Navigating Urban Planning and Design Responses

Working with the urban sound realm can be challenging for decision-makers, city planners, and urban designers. It requires thoroughly understanding the relationship between place, sound and urbanism and introducing disciplines and opportunities not necessarily present in more established approaches to sound and noise in urban spaces. The role of public space in achieving inclusive, welcoming and environmentally friendly cities, the health and well-being objectives, and the paradigm of spatial consolidation driven by sustainability goals is currently engaging many policymakers and urban design professionals at different levels of the planning and design process and also prompting interdisciplinary design teams to increase their efforts in

Fig. 7.5 The Sonic Gathering Place is a permanent sound art installation in Melbourne, created by Jordan Lacey, that combines biophilic design and field recordings of natural sounds. It also hosts public discussions and presentations on sound and urbanism, as seen at the opening with a performance by the Khyaal ensemble. Photo credit: Tobias Titz

exploring new tools and formats for working with sound in the urban environment (Anderson 2022).

Urban sound is influenced by policy design, noise action planning, soundscaping, walking and listening, pilots and design scenarios, acoustic modelling and Virtual Reality simulations, and other avenues not always clearly linked to the public sound realm. The complexity of urbanism and environmental design in rapidly changing urban spaces rarely allows the many disciplines to derive design responses directly from these tools and working formats alone. In juxtaposing the heterogeneous connections to urban sound, this chapter aims to draw the attention of planners and developers to approach the urban sound realm not from a singular path but from different, sometimes overlapping and ambivalent ones. These paths point to sound-related knowledge in specific contexts where various disciplines explore socio-spatial and physical connections to sound. Combining these fragmentary connections into coherent approaches highly depends on the design teams' knowledge of sound and ability to acquire and integrate it into their practice (Maag et al. 2023).

This process of shaping the urban environment has measurable consequences in terms of acoustic metrics, such as decibels and speech intelligibility, and how citizens experience and value the sound environment, such as ranking it higher or lower and responding to it through their daily routines. Ideally, the ears of planners and developers are trained and experienced in recognising and addressing the acoustic

Fig. 7.6 Continuous Drift is an interactive sound installation in Meeting House Square in Dublin. Initially developed in 2015 by the artist Sven Anderson, working in close collaboration with Dublin City Council, this permanent public artwork features contributions from over 30 artists and collectives. Photo credit: Ros Kavanagh

implications of their plans. In practice, however, considerations of the urban sound realm are limited to specific aspects, each addressing only a fragment of planned and designed sound quality. Even when sound is not on the planning agenda, urban designers, planners, developers, and others involved in city-making processes have a particular responsibility, through their disciplines, for shaping the urban sound realm. This responsibility requires interdisciplinary collaborations across various fields, including environmental noise and urban sound, architecture, environmental design and urbanism.

Planners and developers committed to a more versatile approach to urban sound are training their ears, so to speak, by gaining experience and transferring knowledge on a project-by-project basis. This makes integrating sound into the many objectives specified in their daily planning agenda easier. Such an agenda may start with a specific programme, such as city climate strategies, noise action plans, or Sustainable Development Goals explored above, which planners and developers can push forward into more sound-related project agendas (Maag 2013). For each project stage, they will be able to identify the relevant sound-related objectives and adjust their design and planning methods and protocols to address these objectives efficiently. At the same time, they can learn to remain open-minded about framing and achieving these objectives by involving disciplines relevant to the project, such as architecture, engineering, environmental design, and public art, and by considering impulses from public engagement activities.

Urban sound qualities can be articulated by a well-guided and stable planning and design process, as well as by providing sufficient everyday spaces for people to interact with the environment and engage in social routines to explore the urban sound realm. This balanced cultivation of urban sound is crucial for cities home to an increasing number of people with different backgrounds and expectations. By learning how to frame sound-related objectives through the planning and design process, planners and developers will create places that are easier and more enjoyable for citizens to engage with and get their ears around, thus enhancing the urban sound experience.

Acknowledgements We are grateful to all those who have articulated our text through careful feedback and stimulating comments. Special thanks go to the team of editors for extensive reviews and to Sven Anderson, Andres Bosshard, Lex Brown and Jordan Lacey for engaging in discussions about working with sound in cities.

References

AFRY, Efterklang (2020) Ljudmätning av ljudmuren 78250402. Helsingborg

Aletta F, Kang J, Axelsson Ö (2016) Soundscape descriptors and a conceptual framework for developing predictive soundscape models. Landsc Urban Plan 149:65–74. https://doi.org/10.1016/j.landurbplan.2016.02.001

Aletta F, Oberman T, Mitchell A, Erfanian M, Kang J (2023) Soundscape experience of public spaces in different world regions: a comparison between the European and Chinese contexts via a large-scale on-site survey. J Acoust Soc Am 154:1710–1734. https://doi.org/10.1121/10.0020842

Amphoux P, Jaccoud C, Meier H, Meier-Dallach HP Gehring M, Bardyn J-L, Chelkoff G (1991) Aux écoutes de la ville : la qualité sonore des espaces publics européens, méthode d'analyse comparative. Enquête sur trois villes suisses (report). CRESSON, IREC : Institut de Recherche sur l'Environnement Construit

Anderson S (2018) New strategies for sound in the public realm: Integrating a publicly-controlled sound installation in an active city square. In: Proceedings of inter-noise 2018. Presented at the 47th international congress and exposition on noise control engineering, 26–29 August 2018 Chicago, INCE-USA, pp 4989–5000

Anderson S (2022) Beyond standards: in search of heterogeneous approaches to sound in the design and planning of the public realm. In: Proceedings of inter-noise 2022. Presented at the 51th international congress and exposition on noise control engineering, 21–24 August 2022 Glasgow, INCE-USA, pp 2561–2567

Axelsson Ö, Nilsson ME, Berglund B (2010) A principal components model of soundscape perception. J Acoust Soc Am 128:2836–2846. https://doi.org/10.1121/1.3493436

Axelsson Ö (2015) How to measure soundscape quality. In: Proceedings of EuroNoise 2015. Presented at the EuroNoise 2015, 31 May - 3 June Maastricht, Maastricht, pp 1477–1481

Bartalucci C, Luzzi S (2020) The soundscape in cultural heritage. IOP Conference Series: Materials Science and Engineering 949:012050. https://doi.org/10.1088/1757-899X/949/1/012050

Basner M, Babisch W, Davis A, Brink M, Clark C, Janssen S, Stansfeld S (2014) Auditory and non-auditory effects of noise on health. Lancet 383:1325–1332. https://doi.org/10.1016/S0140-6736(13)61613-X

Bonacker M (2018) Avoiding neighborhood complaints due to of urban construction site noise by information and communication. In: Proceedings of inter-noise 2018. Presented at the 47th international congress and exposition on noise control engineering, 26–29 August 2018 Chicago, INCE-USA, pp 6000–6003

Bonacker M, Bachmeier B (2018) Kommunale Praxis der Öffentlichkeitsbeteiligung bei der Lärmaktionsplanung. Lärmbekämpfung 13:6–9

Brown AL, Muhar A (2004) An approach to the acoustic design of outdoor space. J Environ Planning Manage 47:827–842

Brown AL, Kang J, Gjestland T (2011) Towards standardization in soundscape preference assessment. Appl Acoust 72:387–392

Brown L (2021) The sonic environment in urban planning, environmental assessment and management. In: Bull M, Cobussen M (eds) The Bloomsbury handbook of sonic methodologies, pp 217–234. https://doi.org/10.5040/9781501338786.ch-012

Brown AL, Van Kamp I (2017) WHO environmental noise guidelines for the European region: a systematic review of transport noise interventions and their impacts on health. Int J Environ Res Public Health 14. https://doi.org/10.3390/ijerph14080873

Catillaz A, Fischer F (2018) Lärmbelastung der Schweiz. Ergebnisse des nationalen Lärmmonitorings sonBASE, Stand 2015. Bundesamt für Umwelt BAFU, Bern

Cerwén G, Kreutzfeldt J, Wingren C (2017) Soundscape actions: a tool for noise treatment based on three workshops in landscape architecture. Front Architect Res 6:504–518. https://doi.org/10.1016/j.foar.2017.10.002

Diekmann A, Bruderer Enzler H, Hartmann J, Kurz K, Liebe U, Preisendörfer P (2023) Environmental inequality in four European cities: A study combining household survey and geo-referenced data. Eur Sociol Rev 39:44–66. https://doi.org/10.1093/esr/jcac028

United Nations (2019) World urbanization prospects: the 2018 revision (No. ST/ESA/SER.A/420). United Nations, Department of Economic and Social Affairs, Population Division, New York

Echevarria Sanchez GM, Van Renterghem T, Thomas P, Botteldooren D (2016) The effect of street canyon design on traffic noise exposure along roads. Build Environ 97:96–110. https://doi.org/10.1016/j.buildenv.2015.11.033

Echevarria Sanchez GM, Van Renterghem T, Sun K, De Coensel B, Botteldooren D (2017) Using Virtual Reality for assessing the role of noise in the audio-visual design of an urban public space. Landsc Urban Plan 167:98–107. https://doi.org/10.1016/j.landurbplan.2017.05.018

Eggenschwiler K, Heutschi K, Taghipour A, Pieren R, Gisladottir A, Schäffer B (2022) Urban design of inner courtyards and road traffic noise: influence of façade characteristics and building orientation on perceived noise annoyance. Build Environ 224:109526. https://doi.org/10.1016/j.buildenv.2022.109526

Estévez Mauriz L, Forssén J, Kropp W, Zachos G (2016) Urban space and the sound environment: Transport system, urban morphology, quiet side and space users in the SONORUS project. In: Proceedings of inter-noise 2016. Presented at the 45th international congress and exposition on noise control engineering, 21–24 August 2016 Hamburg, DEGA, Berlin, pp 2928–2938

European Commission (2018) Sustainable development in the European Union – Monitoring report on progress towards the SDGs in an EU context – 2018 edition. European Commission, Brussels

European Environment Agency (2014) Good practice guide on quiet areas (EEA Technical report No. No 4/2014). European Environment Agency, Copenhagen

European Environment Agency (2016) Quiet areas in Europe – The environment unaffected by noise pollution (EEA report no. 14/2016). European Environment Agency, Copenhagen

European Environment Agency (2019) Environmental noise in Europe – 2020 (EEA Report No. 22/2019). European Environment Agency, Copenhagen

European Environment Agency (2022) Outlook to 2030 – Can the number of people affected by transport noise be cut by 30%? (EEA Briefing No. No 13/2022). European Environment Agency, Copenhagen

European Parliament and the Council of the EU (2002) Directive 2002/49/EC of the European Parliament and of the Council of 25 June 2002 relating to the assessment and management of environmental noise

FAMOS (2022) FActors MOderating people's Subjective reactions to noise. Project Report (No. FAMOS Deliverable D4.6). Conference of European Directors of Roads

Swiss Federal Council (2022) Botschaft zur Änderung des Umweltschutzgesetzes

Fink D (2019) A new definition of noise: Noise is unwanted and/or harmful sound. Noise is the new 'secondhand smoke'. In: Proceedings of meetings on acoustics, 39:050002. https://doi.org/10.1121/2.0001186

Fryd J, Nøhr Michelsen L, Bendtsen H, Iversen LM, Holm Petersen T (2016) Noise annoyance from urban roads and motorways. Survey of the noise annoyance experienced from road traffic for residents along motorways and urban roads (No. 565–2016). Vejdirektoratet, Copenhagen

Gidlöf-Gunnarsson A, Öhrström E (2007) Noise and well-being in urban residential environments: the potential role of perceived availability to nearby green areas. Landsc Urban Plan 83:115–126. https://doi.org/10.1016/j.landurbplan.2007.03.003

Gisladottir A (2021) Transferring knowledge of urban sound – Informing the urban planning and design practice with sound-aware expertise (thesis). Aarhus University, Aarhus, Graduate School of Technical Sciences

Gisladottir A, Kirkegaard PH, Maag T, Holst Laursen LL (2018) Key elements related to context and morphology for the acoustic design of urban environments. In: Proceedings of inter-noise 2018. Presented at the 47th international congress and exposition on noise control engineering, 26–29 August 2018 Chicago, INCE-USA, pp 3868–3876.

Hallin A, Halling C, Lindqvist M, Åkerlöf L (2006) Trafikbuller och planering 3 – Ljudkvalitets-poäng. Länsstyrelsen i Stockholms län, Stockholm

Han X, Huang X, Liang H, Ma S, Gong J (2018) Analysis of the relationships between environmental noise and urban morphology. Environ Pollut 233:755–763. https://doi.org/10.1016/j.envpol.2017.10.126

Haselhoff T, Lawrence B, Hornberg J, Ahmed S, Sutcliffe R, Gruehn D, Moebus S (2022) The acoustic quality and health in urban environments (SALVE) project: study design, rationale and methodology. Appl Acoust 188:108538. https://doi.org/10.1016/j.apacoust.2021.108538

Heinrichs E, Leben J, Cancik P (2018) Ruhige Gebiete. Eine Fachbroschüre für die Lärmaktions-planung. Umweltbundesamt, Dessau-Rosslau

Hornikx M (2016) Ten questions concerning computational urban acoustics. Build Environ. https://doi.org/10.1016/j.buildenv.2016.06.028

HOSANNA (2013) Novel solutions for quieter and greener cities (EU Seventh framework programme, theme 7, sustainable surface transport). Chalmers University of Technology, Gothenburg

International Organization for Standardization (2014) ISO 12913–1. Acoustics – soundscape – part 1: definition and conceptual framework

Irvine KN, Devine-Wright P, Payne SR, Fuller RA, Painter B, Gaston KJ (2009) Green space, soundscape and urban sustainability: an interdisciplinary, empirical study. Local Environ 14:155–172. https://doi.org/10.1080/13549830802522061

Kang J, Aletta F, Gjestland TT, Brown LA, Botteldooren D, Schulte-Fortkamp B, Lercher P, van Kamp I, Genuit K, Fiebig A, Bento Coelho JL, Maffei L, Lavia L (2016) Ten questions on the soundscapes of the built environment. Build Environ 108:284–294. https://doi.org/10.1016/j.buildenv.2016.08.011

Kaplan S (1995) The restorative benefits of nature: toward an integrative framework. J Environ Psychol 15:169–182. https://doi.org/10.1016/0272-4944(95)90001-2

King EA (2022) Here, there, and everywhere: how the SDGs must include noise pollution in their development challenges. Environ: Sci Policy Sustain Develop 64:17–32. https://doi.org/10.1080/00139157.2022.2046456

Krogh Groth S, Mansell J (Eds) Negotiating noise, Publications from the Sound Environment Centre at Lund University. Sound Environment Centre at Lund University, Lund. https://doi.org/10.37852/oblu.115

Kropp W, Forssén J, Mauriz LE (2016) Urban sound planning – the SONORUS project. Chalmers University of Technology, Gothenburg

LaBelle B (2019) Acoustic territories: Sound culture and everyday life (2nd edn). Continuum Books/Bloomsbury, New York

Lacey J (2022) Three tools for sonic rupture: translating ambiance, biophilic sound design and more-than-human listening. Loci Communes 1–24. https://doi.org/10.31261/LC.2022.02.02

Leereveld T, Margaritis E (2023). Comparison of predicted and perceived tranquility in woonerf streets: a case study of Groningen and Leeuwarden. https://doi.org/10.2139/ssrn.4436203

Leiper A, Hood A (2022) Noise and deprivation in Scotland's four largest cities: Glasgow, Edinburgh, Aberdeen and Dundee. In: Proceedings of inter-noise 2022. Presented at the 51th international congress and exposition on noise control engineering, 21–24 August 2022 Glasgow, INCE-USA, pp 2980–2989

Lynch K (1960) The image of the city. MIT Press, Cambridge

Maag T (2013) Cultivating urban sound – Unknown potentials for urbanism (thesis). Oslo School of Architecture and Design.

Maag T (2021) Encouraging resident-grown urban sound quality. In: Krogh Groth S, Mansell J (Eds) Negotiating noise, Publications from the Sound Environment Centre at Lund University. Sound Environment Centre at Lund University, Lund. https://doi.org/10.37852/oblu.115

Maag T (2016) The quiet city – planning and designing public urban spaces that meet people's needs. In: Proceedings of inter-noise 2016. Presented at the 45th international congress and exposition on noise control engineering, 21–24 August 2016 Hamburg, DEGA, Berlin, pp 3935–3940

Maag T, Bosshard A, Anderson S (2019) Developing sound-aware cities: a model for implementing sound quality objectives within urban design and planning processes. Cities Health 0:1–15. https://doi.org/10.1080/23748834.2019.1624332

Maag T, Anderson S, Gisladottir A (2023) Transferring sound-related expertise into urban planning and design practice on a project-by-project basis. In: Proceedings of inter-noise 2023. Presented at the 47th international congress and exposition on noise control engineering, 20–23 August 2023 Chiba/Tokyo

Maffei L, Brambilla G, Di Gabriele M (2016) Soundscape as part of the cultural heritage. In: Kang J, Schulte-Fortkamp B (eds) Soundscape and the built environment. CRC Press

Margaritis E, Kang J (2016) Relationship between urban green spaces and other features of urban morphology with traffic noise distribution. Urban Forest Urban Greening 15:174–185. https://doi.org/10.1016/j.ufug.2015.12.009

Margaritis E, Kang J (2017) Relationship between green space-related morphology and noise pollution. Ecol Ind 72:921–933. https://doi.org/10.1016/j.ecolind.2016.09.032

Ouzounian, G, 2014. Acoustic mapping: Notes from the interface. In: Gandy, M, Nilsen, B. (Eds.), The Acoustic City. Jovis, Berlin.

Swiss Parliament (2018) Motion Flach Beat. Siedlungsentwicklung nach innen nicht durch unflexible Lärmmessmethoden behindern

Payne SR, Bruce N (2019) Exploring the relationship between urban quiet areas and perceived restorative benefits. Int J Environ Res Public Health 16:1611. https://doi.org/10.3390/ijerph16091611

Pheasant RJ, Fisher MN, Watts GR, Whitaker DJ, Horoshenkov KV (2010) The importance of auditory-visual interaction in the construction of 'tranquil space.' J Environ Psychol 30:501–509. https://doi.org/10.1016/j.jenvp.2010.03.006

Pont MB, Forssén J, Haeger-Eugensson M, Gustafson A, Rosholm N (2023) Using urban form to increase the capacity of cities to manage noise and air quality. Urban Morphol 27:51–69. https://doi.org/10.51347/UM27.0003

Quadmap (2015) Guidelines for the identification, selection, analysis and management of quiet urban areas (No. LIFE10 ENV/IT/000407). QUADMAP QUiet Areas Definition and Management in Action Plans

Radicchi A (2017) A pocket guide to soundwalking. Some introductory notes on its origin, established methods and four experimental variations. In: Besecke A, Meier J, Pätzold R, Thomaier S (eds) Perspectives on urban economics – A general merchandise store. Universitätsverlag der TU Berlin, Berlin, pp 70–73

Radicchi A (2018) From crowdsourced data to open source planning: the implementation of the Hush City App in Berlin. In: Proceedings of inter-noise 2018. Presented at the 47th international congress and exposition on noise control engineering, 26–29 August 2018 Chicago, INCE-USA, pp 3747–3754

Schäffer B, Brink M, Schlatter F, Vienneau D, Wunderli JM (2020) Residential green is associated with reduced annoyance to road traffic and railway noise but increased annoyance to aircraft noise exposure. Environ Int 143:105885. https://doi.org/10.1016/j.envint.2020.105885

Senatsverwaltung für Umwelt, Verkehr und Klimaschutz (2020) Lärmaktionsplan Berlin 2019–2023. Anlage 10: Ruhige Gebiete und städtische Ruhe- und Erholungsräume. Senatsverwaltung für Umwelt, Verkehr und Klimaschutz, Berlin

Southworth M (1969) The sonic environment of cities. Environ Behav 1:49–70

Steele D (2018) Bridging the gap from soundscape research to urban planning and design practice: How do professionals conceptualize, work with, and seek information about sound? (thesis). McGill University, Montreal

Suter B, Peter M, Sturm U, Ruf S, Egger C (2014) Akzeptanz der Dichte. Amt für Raumentwicklung Kanton Zürich, Zürich

Taghipour A, Sievers T, Eggenschwiler K (2019) Acoustic comfort in virtual inner yards with various building facades. Int J Environ Res Public Health 16:249. https://doi.org/10.3390/ijerph16020249

Taylor EJ, Hurley J (2016) Not a lot of people read the stuff: Australian urban research in planning practice. Urban Policy Res 34:116–131. https://doi.org/10.1080/08111146.2014.994741

United Nations General Assembly (2015) Transforming our world: The 2030 agenda for sustainable development. United Nations General Assembly

Van Kamp I, Klæboe R, Brown AL, Lercher P (2016) Soundscapes, human restoration, and quality of life. In: Kang J, Schulte-Fortkamp B (eds) Soundscape and the built environment

Van Renterghem T (2019) Towards explaining the positive effect of vegetation on the perception of environmental noise. In: Urban forestry and urban greening, urban green infrastructure – connecting people and nature for sustainable cities, vol 40, pp 133–144. https://doi.org/10.1016/j.ufug.2018.03.007

Van Renterghem T, Botteldooren D (2010) The importance of roof shape for road traffic noise shielding in the urban environment. J Sound Vib 329:1422–1434. https://doi.org/10.1016/j.jsv.2009.11.011

Van Renterghem T, Hornikx M, Forssén J, Botteldooren D (2013) The potential of building envelope greening to achieve quietness. Build Environ 61:34–44. https://doi.org/10.1016/j.buildenv.2012.12.001

Vienneau D, de Hoogh K, Faeh D, Kaufmann M, Wunderli JM, Röösli M (2017) More than clean air and tranquillity: residential green is independently associated with decreasing mortality. Environ Int 108:176–184. https://doi.org/10.1016/j.envint.2017.08.012

World Health Organization (2011) Burden of disease from environmental noise – Quantification of healthy life years lost in Europe. The WHO Regional Office for Europe, Copenhagen

World Health Organization (2018) Environmental noise guidelines for the European region. The WHO Regional Office for Europe, Copenhagen

Yanaky R, Tyler D, Guastavino C (2023) City Ditty: An immersive soundscape sketchpad for professionals of the built environment. Appl Sci 13:1611. https://doi.org/10.3390/app13031611

Chapter 8
Governance

Benjamin Fenech and Natalie Riedel

Abstract Governance refers to those regulatory processes, mechanisms and organizations through which political actors influence environmental actions and outcomes. Governance processes, mechanisms and organizations relevant to sound and health can vary greatly between continents, countries and even regions. In this chapter we ·present some examples of governance directly relevant to sound and health, from the international to the national level. We then discuss a common shortcoming of current approaches, i.e. that social variations of environmental quality and health are rarely taken into account. We propose the behaviour change wheel as a tool to identify actions that different political actors can take to ensure more equitable outcomes.

> We all want quiet. We all want beauty… We all need space.
> Unless we have it, we cannot reach that sense of quiet in
> which whispers of better things come to us gently. (National Trust)
>
> Octavia Hill, 1883
> Co-founder of the National Trust, England

8.1 Preamble

For the purposes of this chapter, we are using the definition of governance by Lemos and Agrawal (2006), i.e.

B. Fenech (✉)
Environmental Hazards and Emergencies Department, UK Health Security Agency, London, UK
e-mail: Benjamin.Fenech@ukhsa.gov.uk

N. Riedel
Department of Spatial Planning, Urban Research Unit, City of Muenster, Münster, Germany
e-mail: natalie.riedel@udo.edu

© The Author(s) 2025
I. van Kamp and F. Woudenberg (eds.), *A Sound Approach to Noise and Health*,
Springer-AAS Acoustics Series, https://doi.org/10.1007/978-981-97-6121-0_8

the set of regulatory processes, mechanisms and organizations through which political actors influence environmental actions and outcomes. Governance is not the same as government. It includes the actions of the state and, in addition, encompasses actors such as communities, businesses, and NGOs.

These regulatory processes, mechanisms and organizations vary across continents, countries and even regions, and this chapter can only provide a limited overview of the subject area. This chapter is somewhat biased towards governance systems within the Europe region. This is partly due to the provenance of both authors, and partly due to an ambitious programme by the World Health Organization Regional Office for Europe and the European Union over the past two decades to push the health effects of noise up the political agenda. The interested reader is encouraged to browse through the many references at the end of this chapter, which cover a much broader geographical and situational scope.

8.2 Introduction

Many factors determine the way in which sound and noise can be 'governed'. The physical agent, sound, is an essential ingredient, or nutrient, to humans and fauna. Sound allows us to communicate and receive essential information on our surroundings. Therefore, the conventional toxicological approach of avoiding harm from a known hazard by eliminating the exposure completely is not appropriate. Sound always has meaning as Rainer Guski explains in Chap. 3, and it can be health enhancing or harmful depending on the context, personal and social preferences and vulnerability. This leads to tensions between equality versus equity approaches to sound governance (Riedel et al. 2022).

Regulations of environmental hazards are often expressed in terms of limits or targets for short and/or long-term exposure. For sound, setting limits is a complex process that needs to be informed by short and/or long-term exposure levels, the severity of associated health outcomes, the strength of those associations, the strength of the evidence, the prevalence of noise in the living environment, the prevalence of the health outcome in the population and the societal consequences of meeting those limits (see Case Study 2 textbox in this chapter). At a population level, noise exposure levels are linked to health effects: the higher the exposure, the stronger the effect. However, some people may be highly annoyed even at low exposure levels because of negative feelings towards the noise or the person or authority responsible. As exposure increases, other factors such as coping capacity and access to quiet (in space or time) may play a more important role in modifying the exposure-health pathway. Achieving improved health outcomes at a population level is therefore likely to require a careful balance of controlling noise exposure and addressing these mediating and moderating non-acoustic factors (Fenech et al. 2021).

Governance measures addressing threats to public health also need to reflect interactions with the wider determinants of health, including potential co-benefits and unintended consequences of regulations and interventions. A review by Peris and Fenech (2020) found that sound and noise interact with six domains in Barton and Grant's Health Map (Barton and Grant 2006): Lifestyle, Community, Local Economy, Activities, Built Environment and Natural Environment (see Fig. 8.1), and they argue that a holistic approach is needed to capture the full impact of sound and noise on public health.

Whilst these considerations present challenges from a regulatory and governance perspective, they also offer opportunities. Defining a clear boundary between sound that is health enhancing and that which is damaging to health is difficult, if not impossible. Conversely, opportunities to protect and enhance human health are not solely dictated by sound levels, but also through the non-acoustic factors that interact with the sound-health pathway. This overarching approach can be visualised by combining the pathogenesis/salutogenesis approach with the wider determinants of health (see Fig. 8.1). To date, most governance measures have focused on the pathogenetic route (reducing the harmful effects of noise). However, interest in the salutogenetic

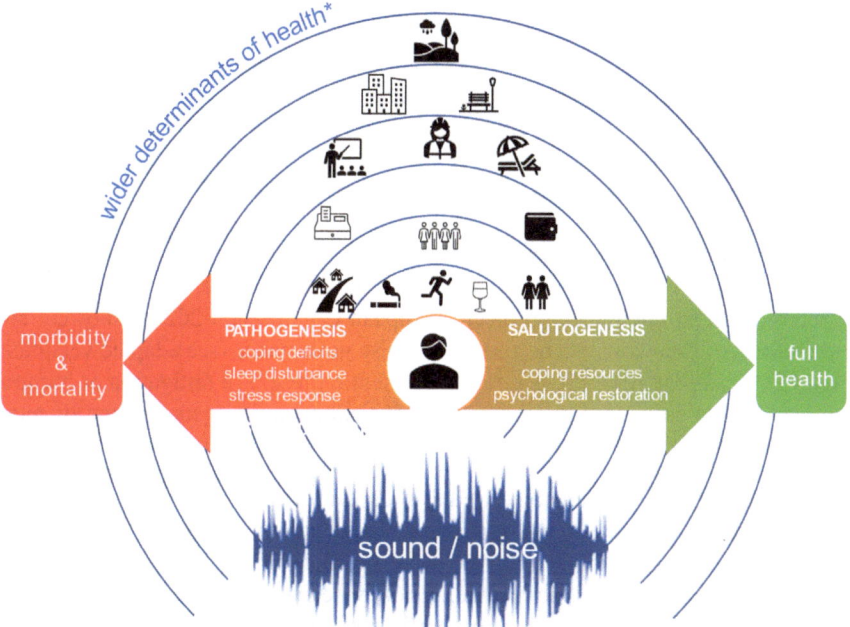

Fig. 8.1 Sound and noise can influence an individual's health both directly but also through complex interactions (Peris and Fenech 2020) with the wider determinants of health: lifestyle, community, local economy, activities, built and natural environment (represented by the icons within concentric circles) as defined by the Barton and Grant model of the determinants of wellbeing and health in our cities (Barton and Grant 2006; Dahlgren and Whitehead 2007). These direct and indirect effects can lead to both positive and negative health outcomes, depending on the situation and the ability to cope with the external stimulus. This figure illustrates the multitude of opportunities and challenges of setting up governance frameworks to deal with sound in our living environment

approach (enhancing the positive effects of sound) is gathering momentum, thanks to a growing interest in soundscapes and health.

8.3 DPSEEA Framework

The DPSEEA (Drivers, Pressures, State, Exposure, Effects and Action) framework (Corvalán et al. 1996) can be used to map the different relationships in governance. Within this framework, the driving forces component (D) refers to the factors that motivate and push the environmental processes involved, including population growth, technological and economic development and policy intervention. These forces result in pressures (P) on the environment, often generated by all sectors of economic activity including energy production, manufacturing, and transport. In response to these pressures, the state of the environment (S) is often modified. The changes involved may be complex and at different geographic scales. When people are exposed to these environmental hazards, then risks to health may occur. Exposure (E_1) thus refers to the intersection between people and the hazards inherent in the environment. Exposure to environmental hazards, in turn, leads to a wide range of health effects (E_2). These may vary in type, intensity and magnitude depending upon the type of hazard to which people have been exposed, the level of exposure and the number of people involved. This framework creates a system that transparently links responses/actions (A) (that require political reforms, investment and buy-in) to the five different components, which in turn facilitate the implementation of decisions that directly address environmental and human health concerns.

The European Phenomena project (European Commission 2021a) showed how the DPSEEA framework can be applied to noise governance (see adapted version in Fig. 8.2), where the response/action (A) element represents a mitigation of harmful noise pollution as well as the adaptation of actions and targets for one or more elements of the DPSEEA framework:

- Driving forces (D) are characterised as social, demographic and economic activities that motivate the relevant process. These can include population growth, urbanisation, increasing mobility (vehicular traffic) of people and goods, technological development, economic and/or policy development, infrastructure development and public opposition to new infrastructure. The response/action to a driver would constitute doing fewer or none of the activities producing noise, for example constraining growth at an airport or designing compact cities that reduce the dependence on motorised vehicles.
- Pressures (P) may include the increase of noise emissions that can have an impact on biological systems and human health. The response to a pressure could be the reduction of noise-producing activities by changing the relevant processes or by making activities less impactful, such as facilitating active travel.
- Environmental State (S) is represented by the spatial distribution of noise. Response to this problem would be to encourage mitigation of harmful noise levels

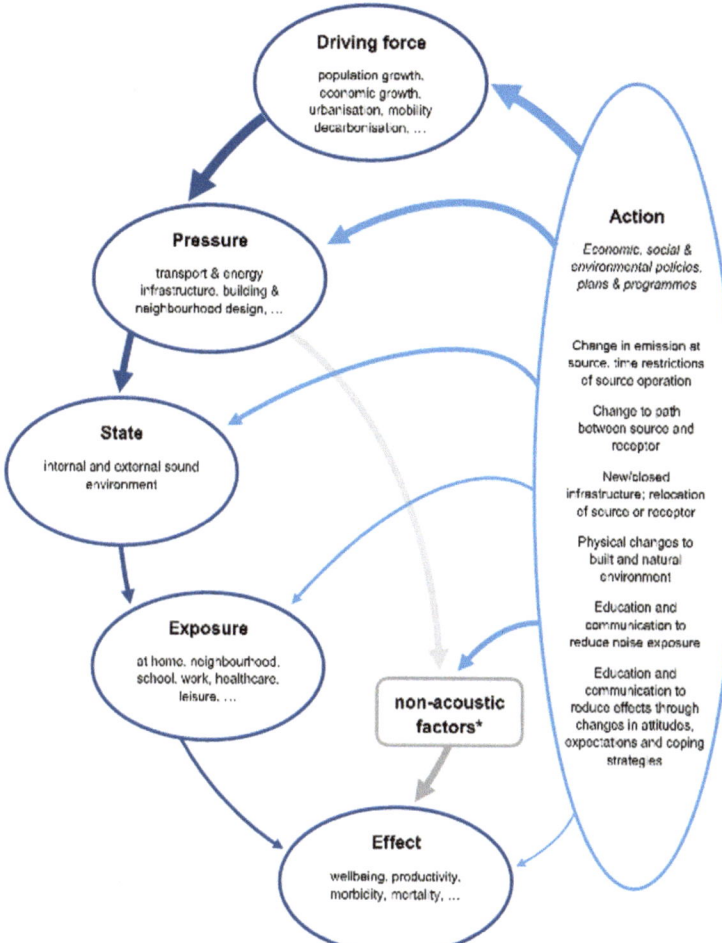

Fig. 8.2 Adapted version of the DPSEEA framework originally conceptualised by Corvalán et al. (1996)

(e.g. by embedding good acoustic design principles in urban master planning (Kropp et al. 2016)).

- Exposure (E_1) refers to the intersection between the number of people exposed to different average noise levels in the environment. Exposure may be quantified by monitoring or modelling techniques.
- Exposure to noise results in adverse health effects (E_2). Impacts can include death and/or illness due to heart disease and stress, sleeping disorder, reduced productivity, cognitive impairment and mental health issues. The response to such effects can be taken by measures at source, in the transmission path or at

the receiver, which aim, essentially, to make the environment (and/or receptors) more resilient to the effects/impacts.

- Actions (A) are responses to effects which may require changes in legislation, policies or hazard management approaches including noise monitoring and control, awareness raising, education, treatment and rehabilitation. Each action should be appropriate and balanced to achieve the desired outcome, aiming to maximise co-benefits and minimise unintentional adverse impacts (e.g. reduction of an increasing traffic vs. mobility needs). While developing appropriate actions to tackle noise pollution, decision-makers should aim to achieve optimal results with their interventions from an environmental, social and economic perspective.
- Non-acoustic factors are personal, social, environmental and/or situational contextual factors that moderate the relationship between Exposure and Health Effects, and are influenced by both Pressures and Actions. For example, an Action to install sound insulation measures in dwellings is more likely to lower annoyance if it is accompanied by good communication and giving citizens a meaningful say in the process (Brown and Van Kamp 2017).

In the next sections, we give an overview of the key "political actors" driving existing sound governance, from the international to the national. We then look at one of the shortcomings of most current approaches, i.e. that social variations of environmental quality and health are rarely taken into account. Finally, we propose the behaviour change wheel as a tool to demonstrate what actions different political actors can take to ensure more equitable outcomes.

8.4 A Global to Local Approach to Sound Governance

The physics of sound propagation means that noise is often a local problem (*local* meaning anything from the adjacent room in a dwelling to tens of kilometres away for elevated and low-frequency sound sources). Personal, local, regional and national beliefs and cultural norms often play an important role in how sound translates into health outcomes. However, this does not mean that governance has to be restricted to a local geographical level. Health-based recommendations, noise emission criteria for cross-boundary vehicles and equipment, and conceptual frameworks for noise mitigation frameworks are all good examples of where governance can be set at the regional, national or international level. Therefore, it is important that anyone interested in resolving a specific noise issue has a good understanding of what governance measures exist at all levels.

The next section provides an overview of governance available at different geographical tiers, with a specific focus on:

- international policies aiming to protect human health and creating sustainable communities;
- international standardisation aimed at creating a level playing field;

- transnational and national legislative frameworks aimed at creating top-down pressure on stakeholders and authorities operating and/or overseeing noise from a source or receiver perspective; some of which may be driven by bottom-up pressure from citizens and stakeholders impacted by noise pollution.

8.4.1 International

Human health is a core principle of sustainable development, and the 2030 Agenda for Sustainable Development (https://sdgs.un.org/goals), adopted by all United Nations Member States in 2015, provides a shared blueprint for peace and prosperity for people and the planet, now and into the future. Of the 17 Sustainable Development Goals (SDGs) in the Agenda, SDG 3 (Good Health and Well-being) and SDG 11 (Sustainable Cities and Communities) are the most directly relevant to sound governance. However, there are wider synergies with other SDGs. SDG 7 (Affordable and clean energy) should include consideration of the community response to sound emissions from renewables such as onshore wind. Noise can lead to inequalities in exposure and health outcomes (SDG 10—Reduced Inequalities), whilst SDG 16 (Peace, Justice and Strong Institutions) is linked to the moderating role of environmental justice, community engagement and trust in authorities for the sound–health relationship.

In its 2022 Frontiers report (UN Environment Programme 2022), the UN Environment Programme drew particular attention to noise pollution and its long-term physical and mental health impacts, along with measures that can be implemented to create positive and restorative soundscapes in urban areas.

The Organisation for Economic Co-operation and Development (OECD) adopted the Recommendation on Strengthening Noise Abatement Policies (OECD/LEGAL/0218) (OECD 2022), which recommends that Member countries undertake a significant improvement in their noise abatement policies by:

- Ensuring a more effective enforcement of existing noise abatement regulations;
- Progressively strengthening noise control regulations, and in particular noise emission limits on products that form important items in international trade, such as motor vehicles and aircraft, along the lines of the conclusions of the OECD Conference on Noise Abatement Policies (1980);
- Complementing existing regulations with incentives designed to promote the production and use of quieter products, such as economic instruments, education and information, product labelling, favourable treatment of quieter products and in-use control of products and vehicles;
- Developing measures to finance noise abatement policies, which would limit pressure on public expenditure;
- Protecting the most exposed members of the population by means such as traffic management, the construction of noise barriers, the insulation of buildings; and preventing the creation of new noise situations by appropriate land use planning, especially in urban areas.

8.4.1.1   World Health Organization

One of the core functions of the World Health Organization (WHO) is to develop guidelines that contain recommendations for clinical practice or public health policy. Recommendations are designed to help end-users make informed decisions on whether, when and how to undertake specific actions with the aim of achieving the best possible individual or collective health outcomes. WHO guidelines are neither standards nor legally binding criteria; instead, they are designed to offer guidance in reducing the health impacts based on expert evaluation of the scientific evidence.

The WHO has been active in the field of noise since the 1970s. The first publication dedicated specifically to noise was Environmental Health Criteria 12 Noise, published under joint sponsorship with the United Nations Environment Programme (United Nations Evironment Programme and World Health Organization 1980). The document identified the following effects of noise: interference with communication, hearing loss, disturbance of sleep, stress, annoyance, and effects on performance. It also included recommended noise exposure limits in terms of $L_{A,eq}$: 45 dB indoors during daytime for good speech intelligibility, and 35 dB in the bedroom at night to protect sleep. The recommended outdoor levels were 55 dB during daytime and 45 dB during night-time. It is striking that the significant body of evidence that has been published since then is suggesting that these quantitative criteria are still largely relevant today.

In 1992 the WHO Regional Office for Europe convened a taskforce to develop guidelines for community noise. The term "community noise" was defined as "noise emitted from all sources (outdoor and indoor) except noise at the industrial workplace. A report by the Karolinska Institute, Sweden was expanded with recommendations on environmental noise assessment and management from a WHO Expert Task Force meeting in 1999. This led to the WHO Guidelines for Community Noise (1999) (CNG) (World Health Organization 1999), which replaced the 1980 criteria. The CNG included a more detailed consideration of impact on vulnerable subgroups, and identified new health outcomes of interest: cardiovascular disease, mental illness and cognitive impairment in children. Guideline values were given for specific health effects and for specific environments, including dwellings (outdoor and indoor), schools, hospitals and parks. The report also included a chapter on noise management "*to maintain low noise exposures, such that human health and well-being are protected*", referring to principles such as the precautionary principle, the polluter pays principle and the prevention principle, although it was acknowledged that this is not always achievable. The chapter described various frameworks and models to contextualise noise management, and many of the principles still apply. For example, the evaluation of noise controls must consider technical, financial, social, health and environmental factors, together with implementation speed and enforceability.

The CNG included summaries of noise policies and situations from across the globe, and whilst many of the specified policies have been superseded or updated, the summaries provide a useful illustration of how noise is managed differently across the globe, and especially the contrast between developed and developing countries.

The WHO CNG shone an international spotlight on the importance of noise as an environmental determinant of health, and the 2500+ citations (as of October 2023, Google Scholar) are a testament to its impact. Whilst the experimental and epidemiological evidence on noise and health has strengthened considerably since the CNG were published, many of the underpinning concepts and recommendations remain relevant and pertinent.

Between 2003 and 2006, the WHO Regional Office for Europe implemented the Night Noise Guideline project, co-sponsored by the European Commission. The final Night Noise Guidelines for the European Region (NNG) were published in 2009 (World Health Organization Regional Office for Europe 2009). As direct evidence concerning the effects of night noise on health was sparse, the guidelines also used indirect evidence: the effects of noise on sleep and the relations between sleep and health. The main recommendation was summarised as follows:

Considering the scientific evidence on the thresholds of night noise exposure indicated by $L_{night,outside}$ as defined in the Environmental Noise Directive (2002/49/EC), an $L_{night,outside}$ of 40 dB should be the target of the night noise guideline (NNG) to protect the public, including the most vulnerable groups such as children, the chronically ill and the elderly. $L_{night,outside}$ value of 55 dB is recommended as an interim target for the countries where the NNG cannot be achieved in the short term for various reasons, and where policy-makers choose to adopt a stepwise approach.

The NNG were expressed in terms of L_{night} (a metric already in use in EU legislation) as it combines the number of events and the maximum sound levels per event. However, it was recognised that different noise metrics could be chosen for different health end points. Long-term effects such as cardiovascular disorders are more correlated with indicators summarizing the acoustic situation over a long time period, such as annually average of night noise level outside at the facade, while instantaneous effects such as awakenings are better correlated with the maximum level per event (L_{Amax}), such as from a vehicle pass-by or aircraft flyover.

At the 5th Ministerial Conference on Environment and Health (Parma, 2010), member states of the WHO European Region requested updated noise guidelines that included not only transportation noise sources but also personal electronic devices and wind turbines, which had not yet been considered in previous guidelines. The Environmental Noise Guidelines for the European Region (ENG) (World Health Organization Regional Office for Europe 2018), published in 2018, were the first noise guidelines to adhere to the requirements of the WHO Handbook for Guideline Development (2014).

Following the publication of WHO's community noise guidelines in 1999 and night noise guidelines for Europe in 2009, these latest guidelines represent the next evolutionary step, taking advantage of the growing diversity and quality standards in this research domain. Comprehensive and robust, and underpinned by evidence, they will serve as a sound basis for action. While these guidelines focus on the WHO European Region and provide policy guidance to Member States that is compatible with the noise indicators used in the EU's Environmental Noise Directive, they still have global relevance. Indeed, a large body of the evidence underpinning the recommendations was derived not only from noise effect studies in Europe but also from research in other parts of the world – mainly in Asia, Australia and the United States of America.

The ENG included as new health outcomes adverse birth outcomes, diabetes, obesity and mental well-being. Guideline recommendation values were set for road, railway, aircraft, wind turbine and leisure sources separately, making them source specific rather than environment specific. The noise level recommendations were accompanied by four guiding principles that provide generic advice and support for the incorporation of recommendations into a policy framework:

- Reduce exposure to noise, while conserving quiet areas;
- Promote interventions to reduce exposure to noise and improve health;
- Coordinate approaches to control noise sources and other environmental health risks;
- Inform and involve communities potentially affected by a change in noise exposure.

In addition to environmental noise, the WHO (Geneva) has also been active in noise-induced deafness and hearing loss. A World Report on Hearing (World Health Organization 2021a), published in 2021, projected that by 2050 nearly 2.5 billion people will have some degree of hearing loss and at least 700 million will require hearing rehabilitation. The WHO estimates that unaddressed hearing loss poses an annual global cost of US\$ 980 billion, with 57% of these costs attributed to low- and middle-income countries. Sound is one of the risk factors for hearing loss. In developed countries, long-term sound from environmental sources does not tend to reach levels that permanently damage hearing, but this may be the case in developing countries. Damaging sound levels are more likely to occur in certain occupational settings, and many countries have some form of industrial noise exposure limits in their regulations and recommended practices (United Nations Environment Programme and World Health Organization 1980). Another source of risky exposure is through the leisure and entertainment sectors (clubs, concerts and personal media devices). The WHO estimates that over 1 billion young adults are at risk of permanent, avoidable hearing loss due to unsafe listening practices. A review on hearing loss due to recreational exposure to loud sounds informed the WHO Make Listening Safe Initiative (World Health Organization 2021b). WHO has published standards that outline safe listening features for a variety of situations where unsafe practices are common, including the WHO-ITU Global standard for Safe listening devices and systems (World Health Organization and International Telecommunication Union 2019) and the Global standard for safe listening venues and events (World Health Organization 2022).

8.4.1.2 ICAO—Aviation

The International Civil Aviation Organization (ICAO) is a specialized agency of the United Nations funded and directed by 193 national governments. One of its functions is to research new air transport policies and standardization innovations as directed and endorsed by governments. Industry and civil society groups, and other concerned regional and international organizations, also participate in the exploration and development of new standards at ICAO in their capacity as 'Invited Organizations'. As new priorities are identified by these stakeholders, the ICAO secretariat convenes panels, task forces, conferences and seminars to explore their technical, political and socio-economic aspects, and develop new international standards and recommended practices for civil aviation.

ICAO's Annex 16 to the Chicago Convention, adopted in 1971, focuses on aircraft noise, engine emissions, CO_2 emissions and carbon offset schemes (CORSIA). The document also contains technical manuals dedicated to each topic which describe procedures for the certification of aircraft related to these emission types. The scope of Annex 16 has significantly expanded in the past 50 years to include various new kinds of aircraft (e.g. light propellers and helicopters) and to introduce more stringent emission limit requirements. This information is a critical component for the generation of noise maps around airports.

Another important ICAO policy that is essential for noise governance around airports is the Balanced Approach to Aircraft Noise Management (https://www.icao.int/environmental-protection/pages/noise.aspx). The Balanced Approach consists of identifying the noise problem at a specific airport and analysing various measures available to reduce noise which can be classified into four principal elements:

1. Reduction of noise at source;
2. Land use planning and management;
3. Noise abatement operational procedures;
4. Operating restrictions.

The goal is to address noise problems on an individual airport basis and to identify the noise-related measures that achieve maximum environmental benefit most cost-effectively using objective and measurable criteria.

Guidance and standards are informed by research into topics such as emerging noise reduction technologies, noise impacts from new aircraft concepts (e.g. unmanned air vehicles), and the development of standards and recommended practices for future supersonic aeroplanes. ICAO is also working on the environmental aspects of airport land-use planning, and good practices on airport community engagement. The latter topics have become increasingly important because of global initiatives for airspace modernization, such as the deployment of Performance Based Navigation (PBN). PBN can improve the management of air traffic to ensure that the capacity demands of the flying public can continue to be met. PBN can also provide opportunities to mitigate the noise impacts associated with aircraft operations, however, there are challenges associated with this. Noise demands can sometimes conflict with requirements to reduce air pollution and CO_2 emissions. PBN also alters the spatial community noise exposure—for example by concentrating more flights within a narrow flight path, or by exposing new communities to noise for a more equitable distribution. Evaluating the impact of such changes requires a holistic approach that takes into account noise exposure metrics and relevant non-acoustic factors. The research project ANIMA has investigated exactly this concept, and makes the case for non-acoustic factors to become the fifth pillar of the Balanced

Approach. The final report *Aviation Noise Impact Management* provides a comprehensive overview of the challenges and opportunities in this area (Leylekian et al. 2022).

8.4.1.3   UNECE—Road Vehicles

The United Nations Economic Commission for Europe (UNECE) is one of five regional commissions of the United Nations promoting Pan-European economic integration. It includes 56 member States in Europe, North America and Asia, but all interested United Nations member states may participate in the work of UNECE. Throughout its history, it has introduced a number of regulations relevant to noise, mostly led by the Working Party on Noise and Tyres (Groupe Rapporteur Bruit et Pneumatiques—GRBP) and the Working Party on Railway Transport (SC2). GRBP is a subsidiary body of the World Forum for Harmonization of Vehicle Regulations (WP.29).

In 2015 the UNECE WP.29 adopted a comprehensive revision of UN Regulation No. 51 on vehicle noise (https://unece.org/press/unece-world-forum-harmoniza tion-vehicleregulations-tightens-vehicles-noise-limits-and-adopts). The 2015 revision introduced reduced noise limits for vehicles (passenger cars, light commercial vehicles, light and heavy trucks and buses) to enter into force in three steps from 2016 to 2024. UNECE argues that this regulation has led to a significant gradual reduction in vehicle noise levels from when it originally came into force in 1982.

In 2016 WP.29 adopted a new Regulation on Quiet Road Transport Vehicles (QRTV) which aims to minimize the risk posed by silent cars, without creating a disturbing level of traffic noise (https://unece.org/press/new-un-regulation-keeps-sil ent-cars-becomingdangerous-cars). Over the years pedestrians and other road users have come to rely on the noise emitted by combustion engine vehicles to provide useful information such as the presence of one or more cars, their approximate speed, whether the vehicle is accelerating or decelerating, and so on. These sounds are especially important for the visually impaired. The new Regulation stipulates that quiet cars should be equipped with an Acoustic Vehicle Alerting System (AVAS) to create artificial noise in the speed range from 0 to 20 km/h (for the effects of AVAS see Chap. 10 on future developments). Above 20 km/h, the noise of tyres on the road and the wind noise are audible even from a fully electric car thereby negating the need for a warning system. The Regulation also specifies a maximum overall sound limit to limit noise pollution.

In 2022 WP.29 adopted a new UN Regulation on audible reverse warning (https://unece.org/sustainable-development/press/new-un-regulationharmo nizes-reverse-warning-sound-vehicles). Reverse warning sound is widely used to ensure the safety of people around medium- and heavy-duty vehicles. However such sounds can be problematic in compact neighbourhoods where dwellings are in close proximity to shops and other commercial establishments, and for long-term construction sites in urban areas. The sounds can be especially problematic for late evening/ night-time/early morning operations. The new regulation aims to strike a balance between ensuring a sufficiently recognizable warning sound and avoiding inappropriate noise nuisances for the environment and people. It includes provisions for traditional reverse warning devices with a fixed volume as well as for adjustable devices that automatically select an appropriate volume depending on the ambient noise. In addition, when the same level of safety may be ensured by means of other safety devices (e.g. monitoring systems with reverse cameras are installed on the vehicle and active), the reverse warning sound may temporarily stop.

8.4.1.4 ISO and IEC

International standards aim to bring economic benefit (by removing trade barriers and creating a level playing field) whilst protecting the health of the planet and people. The International Organization for Standardization (ISO) is an independent, non-governmental international organization with a membership of 167 national standards bodies. The International Electrotechnical Commission (IEC) is the world's leading organization for the preparation and publication of international standards for all electrical, electronic and related technologies. Standard development within both organizations is led by technical committees (TC), numbered according to the order they were created. The most relevant technical committee in ISO for sound governance is ISO/TC 43, responsible for standardization in the field of acoustics, including methods of measuring acoustical phenomena, their generation, transmission and reception, and all aspects of their effects on man and his environment. It specifically excludes electro-acoustics and the implementation of specifications of the characteristics of instruments for measuring sound, which falls under the remit of IEC TC 29.

Standardisation also takes place at the supranational and national levels, and these three layers are often inter-linked. For example, in the U.S., the Acoustical Society of America, an international scientific society founded in 1929, develops its own acoustics standards (as an ANSI-accredited Standards Developing Organisation) but also provides U.S. stakeholders with access to international standards development in the ISO and IEC. CEN, the European Committee for Standardization, has a technical committee responsible for the standardisation of all applications for heavy rail and urban rail systems in Europe; some of which are directly relevant to noise emissions.

The majority of standards produced by all these bodies concern the specification and calibration of equipment; and the measurement and/or prediction of sound emission, immission and propagation. Whilst such standards are an important component of sound & health governance, they are beyond the scope of this chapter. However, three particular standards deserve specific mention from a governance perspective.

Long-term noise annoyance normally constitutes a significant proportion of the total disease burden attributable to noise (Jephcote et al. 2023). Given that noise annoyance does not have an ICD9 or ICD10 code, it is important that epidemiological studies quantifying annoyance use a consistent approach (Van Kamp et al. 2018). In 1993, the International Commission on Biological Effects of Noise (ICBEN) began formalizing a standardized methodology for assessing noise annoyance which resulted in reporting guidelines and recommendations. These were later published as a Technical Specification (TS) by the International Standards Organization (ISO) as *ISO/TS 15666 Acoustics—Assessment of noise annoyance by means of social and socio-acoustic surveys* (https://www.iso.org/standard/74048.html). Clark et al. (2021) provide a comprehensive review of the evolvement of this TS over two decades (2001–2021). They also explain the importance of understanding the philosophy and detail that lies behind the specification, as well as following the recommendations in the latest revision of the TS (including the Annexes). Also worth noting is the ongoing debate on whether the two noise reaction questions in the TS are sufficient, on their own, to describe the multi-faceted complex concept that is annoyance (Schreckenberg et al. 2018). Most other health and psychological constructs are assessed using multi-item scales that demonstrate robust psychometric properties when the items are considered together. To address this, Schreckenberg et al. developed a Multiple Item Annoyance Scale (MIAS) that includes the three components of annoyance: noise-related disturbances; affective evaluations and attitudes; and perceptions of control and coping capacity. Such additional components were intended to complement, rather than replace the ISO/TS 15666 Questions. This approach could help to further explain non-acoustic factors and to assess the impact of mitigation or interventions. In 2022, ISO formed a new Working Group (ISO/TC 43/SC 1/WG 68) to develop a new Technical Specification on non-acoustic factors (https://www.iso.org/standard/84809.html).

Whilst a lot of noise and health research and guidance focuses on transportation noise, other sources are equally important. Neighbour noise is an important source and requires a somewhat different governance approach than transport. Noise "control" at the source is less appropriate (although education on the principles of tolerance and respect is important). Instead, a fundamental element in the governance of neighbour noise is the dwelling construction. Building acoustic regulations for housing exists in most countries in Europe and in many countries worldwide. However, complying with such requirements does not guarantee satisfactory conditions for the occupants. *ISO/TS 19488:2021 Acoustics—Acoustic classification of dwellings* (https://www.iso.org/standard/77742.html) is a classification guideline that specifies criteria for six classes (A–F) for dwellings, class A being the highest class and F the lowest class. The purpose of this document is to make it easier for developers to specify a classified level of acoustic quality (sound insulation towards neighbouring premises and external traffic as well as sound from service equipment) for a dwelling, and help users and builders to be informed about the acoustic conditions and define increased acoustic quality. The document can also be applied as a general tool to characterize the quality of the existing housing stock and includes provisions for classifying the acoustic quality before and after renovation has taken place. Sound insulation and room acoustics internally in a dwelling are not included in the acoustic classes defined. This document does not have a legal status in a country, unless decided by its own authorities. However, it could help national authorities and standardization organizations to develop or revise national building regulations and acoustic classification schemes. The standard was based on evidence gathered during the European COST Action TU0901 (2009–2013) project. Rasmussen and

Machimbarrena (2019) provide an interesting insight into some of the challenges in the process of achieving compromise and consensus for these kinds of standards—both technical and political (such as fears of reactions to change by legislators and developers in countries where acoustic classification schemes or regulations already exist).

The first two decades of the twenty-first century saw a rapid rise in interest in an alternative, more holistic approach to sound—the soundscape approach—also discussed in Chaps. 7 and 9 of the present book. This rapid rise was partially driven by, and was also the driving force for the development of the ISO 12913 suite of standards on soundscape. As of 2023, three documents have been published: *Part 1: Definition and conceptual framework* (https://www.iso.org/standard/52161.html), *Part 2: Data collection and reporting requirements* (https://www.iso.org/standard/75267. html), *Part 3: Data analysis* (https://www.iso.org/standard/69864.html) and a *Part 4: Design and Intervention* is in development. ISO 12913-1 defines the term soundscape as "*an acoustic environment as perceived or experienced and/or understood by a person or people, in context*". Soundscape is a separate concept to the acoustic environment: the former relates to the perceptual outcome of the individuals, the latter relates to the whole set of sources generating sounds in a specific space and its physical implications. As such, acoustic environments are neither positive or negative: instead, the perception and experience of the listener within a specific context elicits positive or negative soundscapes, accordingly. The aim of the standard series is to better harmonise soundscape research such that the evidence can be better compared and potentially aggregated. Soundscape is also slowly filtering into local, regional and national policies. For example, *The Environment (Air Quality and Soundscapes) (Wales) Bill 2023* was passed by the Welsh Parliament in November 2023. The official press release (https://www.gov.wales/new-powers-tackle-air-and-noise-pol lution-will-leadcleaner-healthier-and-greener-future) stated that this bill "*will put onus on Welsh Government to make policies that not only tackle unwanted noise, but also protect sounds that matter to people, like the relaxing calls of birdsong and nature, or the welcoming hum of a vibrant town centre. The soundscapes strategy is in response to emerging science on the impacts of sounds on our health and well-being, as well as that of animals. If passed, Wales will be the first country in the UK to introduce such plans.*"

8.4.2   Transnational—European Union

The European Union (EU)'s noise policy is a good example of a governance framework that has been developed with a clear objective—to reduce the health burden attributable to noise in society. The *Fifth EU Environmental Action Programme* (1993–2000) established an objective that "*no person should be exposed to noise levels which endanger health and quality of life*", but it left it up to the Member States to determine the most appropriate implementation at a national level. The subsequent

Green Paper on '*Future Noise Policy*' (1996) offered a vision for creating a comprehensive EU noise regulatory framework with a focus on noise exposure coming from road, rail, air transport and outdoor equipment. Around the same time, the WHO published its *Guidelines for Community Noise,* and this document had a significant influence on subsequent EU policy and legislative developments on noise.

At an EU level, (Directive 2002/49/EC of the European Parliament and of the Council 2002) on the assessment and management of environmental noise (often referred to as the Environmental Noise Directive, or END) is one of the most important legislative instruments for protecting people's health and well-being from excessive noise pollution caused by road, rail and airport traffic, and large industrial installations. It does this by (1) setting a common approach to avoid, prevent and reduce the harmful effects of environmental noise and (2) providing a basis for developing measures to reduce noise emitted by the major sources. It focuses on four action areas: (a) determining exposure to environmental noise and assessing its health effects; (b) ensuring that information on environmental noise and its effects is made available to the public; (c) preventing and reducing environmental noise; and (d) preserving environmental sound quality in areas where it is good. Two key deliverables are strategic noise maps and associated Noise Action Plans.

The END came into force in 2002, and as of 2023, it has undergone three implementation reviews. These reviews deliver a wealth of information on the challenges and opportunities offered by such ambitious governance frameworks—information that is of interest to a global audience. The first implementation report (Commission to the European Parliament and the Council 2011) emphasised that END brought real benefit, leading to the development of the first coherent management system of environmental noise in all Member States. The second evaluation report (European Commission 2016) addressed questions of effectiveness, efficiency, coherence, relevance and EU-added value. The evaluation found that

- the Directive remained highly relevant, and was coherent in itself and with other relevant EU legislation;
- some progress was made towards a common EU approach, but delays in its implementation meant that the Directive had not yet delivered all its potential added value;
- administrative costs were low (median costs per total inhabitants: €0.15 for noise maps and €0.03 for action plans, every 5 years);
- a cost–benefit analysis showed that implemented action plans, including measures for noise management, delivered a favourable cost–benefit ratio of 1:29;
- the Directive can generate EU-added value by providing a level playing field across the EU in which transport infrastructure operators can compete, and by better informing EU policy-making.

The 2nd implementation review also identified considerable differences in implementation approaches between member states. The administrative level at which implementation takes place (i.e. national, regional, and local) was found to vary between agglomerations (large urban areas), roads, railways and airports. This reflects the fact that the END is implemented under the subsidiarity principle. The END does not set any source-specific limit values ("LVs") at an EU level (see next section on National governance for a review of critical noise values in the EU). A particular problem was identified in respect of the timeliness of the completion of Noise Action Plans (NAPs) in agglomerations. In countries that adopted a decentralised approach, many different actors were involved, which made it difficult to coordinate the development of NAPs in an efficient and timely manner. The quality of consultation responses to the publication of draft NAPs was found to vary. Some competent authorities received little input from relevant stakeholders. Conversely, NGOs that participated in consultations stated that although NAPs often include a summary of the consultation responses, it is often unclear how these responses have been taken into account in NAP finalisation. Another difficulty in agglomerations was that the authorities responsible for developing the NAP (often local authorities) do not have strategic or budgetary decision-making powers to determine whether measures included within NAPs are realistic, feasible and can be funded. This was less of a problem for other sources, such as major railways and major roads, where the responsible authorities for action planning may also have budgetary or decision-making powers.

The END requires strategic noise maps to be produced on a five-year cycle. Four rounds have been reported in 2007, 2012, 2017 and 2023. The second implementation review noted improvements in the second round of mapping over the first round due to better quality and availability of input data (European Commission 2016). There was also recognition that the development of a common noise method for calculating noise exposure (CNOSSOS-EU) should, over time, lead to more comparable data between mapping rounds.

Under the European Green Deal, the EU committed itself to achieving a zero pollution ambition for a toxic-free environment. The 2021 zero pollution action plan (European Commission 2021b) focuses on tackling noise at source by securing proper implementation and, where appropriate, by improving the EU's and global regulatory frameworks on road vehicles and their tyres, railways and aircraft. It also highlighted the need to better integrate countries' noise action plans into sustainable urban mobility plans by expanding the clean public transport network and promoting more active travel. Despite 20 years of implementing the END and other national noise policies and national noise limits, noise exposure has remained rather stable and has not decreased. A specific target of reducing the number of people chronically disturbed by transport noise by 30% by 2030 (vs 2017) was set. The first integrated Zero Pollution Outlook Report (European Commission 2022), published in 2022 by the European Commission Joint Research Centre found that the number of people chronically disturbed by road transport noise is unlikely to decline by more than 19% by 2030 (i.e. well below the 30% reduction target) unless a substantial set of additional measures is taken at national, regional and local level and unless reinforced EU action across relevant sectors delivers significant further reduction in noise pollution (see Case study box—Phenomena project (European Commission 2021a)).

Evaluating the Potential Health Benefits of Noise Abatement Measures at a Supra-national Scale—The Phenomena Project

Noise abatement measures are normally explored and evaluated at a project level or for a specific local setting. However, ambitious objectives for noise impact reductions require a more strategic effort at a national or transnational level. The Phenomena project (European Commission 2021a) was commissioned by the European Commission to define measures capable of delivering significant reductions (20–50%) in the health burden due to environmental noise from roads, railways and aircraft by 2030, and explore how legislation could facilitate the implementation of measures. The project focused on geographical areas exposed to levels above the WHO 2018 noise guidelines.

The Phenomena report contains a comprehensive analysis of practical noise abatement solutions covering road, rail and aircraft noise. It also provides indicative costs of noise solutions, although these may vary significantly depending on continent/country. A number of hypothetical scenarios, featuring both single and combined noise abatement solutions, were assessed and ranked on the basis of their expected health burden reduction and benefit to cost ratio (BCR) over a period of 15 years (2020–2035). For road traffic, the most effective scenario was a combination of more quiet roads, quieter tyres and lower vehicle sound limits (health burden reduction of 16–22%; BCR 0.8–4.6). An alternative combined scenario of speed restrictions, car-free zones, quiet facades and dwelling insulation had a similar health burden reduction (16–20%), but much lower BCR (0.04–0.2). For railway noise the scenario combining smoother wheels and rails led to a health burden reduction of 30–42%, with a corresponding BCR of 2–9. The scenario featuring smoother and quieter vehicles and tracks had a higher health burden reduction (37–52%) but smaller BCR (0.9–3.1). Introducing more barriers and traffic reduction had a 5–10% health burden reduction and a BCR of 0.9–4.5, whilst urban planning and reconstruction and more façade insulation led to an 8% health burden reduction and a BCR of 0.2–0.4. For aircraft noise, the single best solution with respect to health burden was a blanket EU-wide ban on night flights (health burden reduction of 37–60%), however acknowledging it would come at a high cost (benefit to cost ratio 0.1–0.2). The combined scenarios considered were operational changes (improved take-off procedures and dispersion or concentration of flights (HBR 25%, BCR −2.4 to −1.8 (cost saving)); accelerated fleet replacement (HBR 27%, BCR −0.1); and a combination of the two (HBR 45%, BCR −0.2 to −0.1).

It is important to note that the Phenomena project focused on the most effective interventions at EU level, and therefore different health reductions and BCRs are likely to be obtained at regional or local level. One such example is noise barriers for road traffic: the cost-effectiveness of noise barriers tends to be low at a national level, partly due to the high costs, but also because they are not feasible in many urban situations. However, in certain situations along motorways and arterial roads with large number of adjacent dwellings they may be an appropriate solution (combined with other measures). Similarly, for railway noise urban reconstruction including tunnelling, screening by buildings and integrated noise abatement, combined with increased facade and building insulation, can have large potential at local level, especially when included as part of new development.

Another important point to note is interdependencies, for example in the case of aviation noise, greenhouse gases and air quality (including NOx and ultrafine particulates). Phenomena also looked at the potential implications of introducing EU-wide noise exposure limits. The estimated health burden reduction and BCR for limits of 60 dB L_{den} and 55 dB L_{night} were: for road traffic noise: 8–19%, BCR of 1–9; for railway noise: 4–8% BCR of 0.2–1.7, for aircraft noise: 14–39% BCR of 0.8–3.1. The project estimated that the maximum technically-feasible noise reduction between 2017 and 2030 is approximately 45%.

The following main drivers were found to support effective/successful implementation of noise policies: complaints and demands from citizens; sufficient funding; initiatives and experience of government authorities; supportive legislation and processes; impact assessments; cooperation among stakeholders; external noise experts; and legally binding noise limits. Conversely, some of the main obstacles to enforceability of noise policies and measures were: financial limitations; lack of competency and initiative; lack of human resources; lack of awareness of noise policies and health impacts of noise abatement measures; competing political priorities; increasing urbanisation; shared responsibilities between multiple authorities; and lack of coherence among legislations.

One of the challenges identified related to the financing of noise interventions, as often these require co-financing by national, regional and/or local authorities, who may have differing long-term strategic priorities. One possible solution is to emphasize the linkage between public health and noise exposure specific to the region or urban area in question.

8.4.3 National

As already discussed, noise governance approaches vary greatly between countries, and sometimes even within the same country. For example, a review for the European Phenomena project (European Commission 2021a) identified a total of 357 different legislations and other instruments in an analysis of transport noise action plans from 22 EU member states. Legislation and policy instruments were mostly established at the national level (59%), followed by the local level (24%) and the regional one (16%), although this varied by country. At the regional and local level, these instruments often address key socio-economic and environmental challenges, such as employment, business development, connectivity/mobility, public services, green transition economy, and governance. Other instruments included urban planning documents, national and regional spatial strategies, transport or mobility plans, environmental strategies, noise abatement guidelines and programmes, transport-specific guidelines, and investment programmes. At the city level, planning, the topic of Chap. 7 of this book, was highlighted as one of the most effective instruments.

Another review, commissioned by the European Network of the Heads of Environment Protection Agencies (EPA Network) and published in 2019, looked at critical noise values in use across 27 European countries, including limit values, relevant noise indicators, and the consequences of exceedance (Peeters and Nusselder 2019). A distinction was made between "noise limits" and "noise targets". For some pollutants, the term "limit value" means a maximum value or an upper limit, that should never be exceeded. However, this appears to be very rare for noise, and for all the countries that reported having limit values, exceptions were possible. Some examples of exceptions were:

- If a dwelling was already there before the legislation came into force;
- If active noise measures are not technically possible, or would require excessive costs (provided that dwellings are noise-insulated to some degree);
- The level may be exceeded (to a certain degree) if the most exposed façade has no openable parts/windows and there is access to a quiet side of the building.

Limits may also refer to a level at which building new infrastructure or new dwellings is no longer allowed. Existing situations with higher noise levels may not be remediated. Many countries also considered a 'limit value' more like a 'target value': a noise level above which noise measures need to be considered, and below which nothing has to be done (see also text box below). The 2nd END implementation review (European Commission 2016) defined limits as values *"whose exceedance generally leads to sanctions, or whose potential exceedance blocks the operation of installations (such as new roads, railways, or industry)"*, and targets as *"values whose exceedance demands the consideration of action to reduce noise"*. The rationale behind national noise limits varied across countries. Nine countries referred to the 1999 WHO guidelines and/or the exposure–response functions used in the WHO 2011 Burden of Disease report, with some choosing values corresponding to a particular annoyance level (between 9 and 15% highly annoyed people). Two countries referred to the END as a basis for their limit values, without further specification. Seven countries set limits on the basis of national studies into local exposure–response relationships, cost/benefit data and/or consultation with the public and/or other stakeholders. Five respondents for countries with noise limits did not answer, or did not know, what was the basis for these values. The consequences of exceeding noise limits also differed across countries. Noise mitigation at source was the most common consequence, occurring in 85–100% of the countries for all noise sources except aircraft noise. Noise mitigation at the receptor (e.g. dwelling insulation) was also common for road, rail and air traffic noise, with several countries explicitly stating that such measures were to be taken if reduction at source was not possible or cost-effective. Prohibition of activities was common (>75%) for wind turbines and industrial activities, but rather uncommon (<20%) for traffic sources. Financial sanctions or compensation was reported for all noise sources, but was more common for air traffic and industry than for road and rail traffic.

Recommendations for Setting National Limit Target Values for Transportation Noise Informed by the 2018 WHO Environmental Noise Guidelines
Adapted from the report **Overview of critical noise values in the European Region** *published by the EPA Network Interest Group on Noise Abatement* (Peeters and Nusselder 2019)

- Legislation should be clear about the objective of any limit or target value, either as a minimum value above which actions should be considered, or as a maximum value that should not be exceeded. A combination of both an upper and a lower value, with room for situational policy in between, is also a possibility.
- The actual significance of a limit value is determined largely by the consequences of exceeding it. When considering new or different values for their noise limits, authorities should regard the legislative system as a whole, including the enforcement and legal consequences. Specifically for existing situations, a trigger to actually assess the noise levels against the limit should exist. This trigger could be national noise maps and action plans.
- When deciding on limit values and their consequences, authorities should be aware that higher levels increase the health impact of noise on the population.
- In order to take into account the WHO-recommended values, noise limits based on indicators different from L_{den}/L_{night} could be used, but their values should be derived using appropriate conversions.
- For transparency and accountability, the rationale behind the actual dB-value should be clear and publicly documented. This should preferably be related to some exposure–response relation, along with cost–benefit and other consideration that may be the basis for that particular value. The WHO-guidelines could provide exposure–response functions for this.
- Following the WHO-recommendations, limit values for road and rail traffic should not be very different and limit values for aircraft should be considerably lower than for road and rail traffic. This consideration is purely from a health perspective, however, and other considerations may apply.

Another dimension of national governance is regulation of housing quality, and in particular sound insulation. Sound insulation between dwellings is arguably the most important physical protection for adverse effects attributable to neighbour noise. Whilst neighbour noise receives far less attention than transportation noise, it represents a significant proportion of community annoyance. For example, analysis from the WHO Large Analysis and Review of European housing and health Status (LARES) (Niemann and Maschke 2004), which combined data from eight city studies in European countries, showed that "neighbour flat" was the second source of noise annoyance (after road traffic). The UK's National Noise Attitudes Survey 2012 (Department for Environment Food and Rural Affairs 2014) found that 11% of the population were "Very or extremely" bothered, annoyed or disturbed by noise from neighbours and/or other people nearby (vs. 8% from road traffic). The European project COST Action TU0901: Integrating and Harmonizing Sound Insulation

Aspects in Sustainable Urban Housing Constructions carried out a comprehensive comparison of sound insulation requirements in 35 European countries (https://www.costtu0901.eu/index.html). Whilst some of the data may now be out of date, the interested reader can find a comprehensive overview of how regulations have evolved in this area in 29 European countries (including Turkey) together with Australia and New Zealand. Despite significant divergence in ideas between international countries, this work led to the development of an ISO Technical Specification for the acoustic classification of dwellings (https://www.iso.org/standard/77742.html).

Whilst the focus in this section has been on Europe, noise regulations and guidance are in place in many other countries and regions. The International Commission on the Biological Effects of Noise (ICBEN) publishes reviews of noise policies from around the world every three years (coinciding with the ICBEN Congresses on Noise as a Public Health Problem (http://www.icben.org/Proceedings.html). Reviews are also published regularly in academic journals. For example, Laplace et al. (2022) review policies and regulations developed in Canada since the 1970s by different levels of government (federal, provincial and municipal governments, as well as other regional and local entities). They concluded that the policy was fragmented, and the large number of stakeholders and measures could be better integrated and harmonized. Perna et al. (2022) document the wide range of specifications observed in road traffic policies across Australia, Europe and North America, including the responsibilities of administrative governments according to the scope (e.g. emission vs. exposure), comparing noise limits by scope and geographic areas and against the WHO's ENG 2018, and comparing measurement protocols across outdoor noise policies. Yokoyama and Kobayashi (2022) investigated the laws and regulations for occupational noise in 27 countries in the Asia–Pacific region.

Inevitably reviews of national policies are bound to lose some of the detail that is present in governance at a national, regional, or sometimes local level. Such reviews are also snapshots in time, and need to be interpreted against a moving landscape of constant iterative policy development and revision. The following two case studies aim to highlight some of the specific approaches from two very different geographical regions (Nepal and Switzerland). These case studies also illustrate that noise governance cannot be based solely on a set of numbers, and the noise and health evidence does not provide all the answers. On one hand, governance needs to take into account a broad range of economic, social and environmental considerations. On the other hand, achieving equitable outcomes requires complementary approaches that go beyond the noise exposure level. For example, using evidence from social justice research, Hauptvogel et al. (2021) develop recommendations on how fairness aspects can be integrated into aircraft noise management to improve the relationship between the airport and its residents, reduce annoyance and enhance the acceptance of local aviation and the airport as a neighbour. This concept is explored in more detail in the last section of this chapter.

Case Study 1—Noise governance in Nepal by Rehana Shrestha, *Associate Researcher at the Department of Social Epidemiology, Institute of Public Health and Nursing Research, University of Bremen*

Many developing countries, including Nepal, have been experiencing rapid, often unmanaged urbanization driven by population growth and rural-to-urban migration of people. A common side effect is the pollution of the environment and degradation of the quality of life of the population living in these urban areas. Noise is a key pollutant linked to industrial, commercial, transport and recreational activities. A report by the Nepal Health Research Council (2003) indicated that average noise levels (L_{den}) in five major urban cities of Nepal were 62 dB for new residential areas, 66 dB for existing residential areas, 73 dB for mixed commercial and residential areas, 69 dB for mixed commercial and tourist areas, and 74 dB for high traffic areas. Studies dating back from the 1980s suggest even higher noise levels in the capital Kathmandu (Shrestha and Shrestha 1985; Manandhar et al. 1987). Few studies have studied the physiological impact of noise, and they have focused generally on occupational noise exposures. High prevalence of noise-induced hearing impairment was reported among certain professions such as traffic personnel, bus drivers, metal and wood industry workers (Ghimire et al. 2019; Robinson et al. 2015; Sanju and Kumar 2016; Shrestha et al. 2011). One study (Joshi et al. 2003) reported that the prevalence of noise-induced hearing loss in the general population living near main roads with noise level above 75 dB(A) was three times higher than of those living in inner urban areas with noise levels below 55 dB(A).

Noise as an environmental hazard has been recognized by the legislation of Nepal for the past two decades. The promulgation of the Environment Protection Act in 1997 (revised in 2019) has led to emission standards for pollution control (GoN 2019). This was followed by the development and endorsement of a National Sound Quality Standard (NASQS) in 2012 that provides sound limit threshold values across various land uses—industrial, commercial, rural residential, urban residential, mixed residential and peace area—and for both day and nighttime (Nepal Gazette 2012). The standard also provides optimum sound emission limits in domains other than public spaces, e.g. for certain household appliances such as water pumps, diesel generators and entertainment goods. Nepal has endorsed additional Acts—the Motor Vehicle and Transport Management Act in 1993, the Labor Act in 1991, and the Industrial Enterprises Act in 1993—with the intention of controlling noise pollution from vehicles, safeguarding of workers from noise impact, and promoting emission-free and environment-friendly industries and enterprises, respectively.

Yet, noise exposure is generally considered a lower priority compared to other environmental problems such as air pollution, water pollution and solid waste management, both in research and policy. After enacting the Environmental Protection Act 1997, the country made some efforts to control noise pollution. For example, the Government declared "No Horn Zones" in sensitive areas like schools and hospitals, since horns are a dominant characteristic of road traffic noise in Nepal's cities. However, such "Horn Restriction Zones" are often not implemented effectively. For example, the Metropolitan Traffic Police Department (MTPD) in collaboration with the Kathmandu Metropolitan City (KMC) re-enforced a "No Horn Regulation" in 2017, which banned all honking except during emergencies, on turning points and for ambulances, fire engines and police vehicles (Malla 2017). However, one study from 2018/2019 (Chauhan et al. 2021) showed that approximately half of the 312 horn events recorded during an observation period of 60 min at 12 different locations were those prohibited by the Regulations. Nevertheless, this noise reduction strategy was effective to some extent in reducing noise in certain zones. Compared to 2015, the noise level reduction (A-weighted) in the year 2018/2019 was found greatest in low traffic zones (7 dB). Reductions of 2 dB were observed in high traffic and residential zones, whereas an increase of 3 dB was observed in commercial areas.

To date, noise assessment studies in Nepal have relied mostly on noise measurements at few selective locations. Comprehensive noise monitoring via nationwide surveys and noise mapping with greater spatial coverage is lacking. Most of the noise-related studies are focused on cities in the Kathmandu valley. Assessment of perceptions of the population towards their own exposure and associated impacts among disadvantaged populations in the context of environmental justice and inequalities are generally less researched. However, recent studies have proven that people-centric approaches and perceptions could play an important role for monitoring noise and health. Choi et al. (2022) found higher noise levels in slums in contrast to non-slums, suggesting high noise annoyance as a good determinant of low quality of life among slum dwellers. Similarly, inequalities in perceived exposure to noise and self-reported health effects were found to be higher in low socio-economic neighborhoods (Flacke et al. 2022).

Given the lack of comprehensive data on noise levels at a higher resolution and the need for more evidence regarding health impacts of noise pollution on various disadvantaged groups of people, participatory sensing approaches can be used to provide a more people-centric, perception-based understanding. Studies have shown how the almost ubiquitous mobile phone can be employed to create a low-cost, open platform to involve the public for measuring, annotating, and localizing noise pollution, thereby informing government officials about the problems, as well as acting as a first step for measuring effectiveness of the interventions that are enforced. Shrestha et al. (2022) demonstrated the feasibility of using digital technology for assessing environmental justice and environmental inequalities in core urban areas and among young people in Nepal.

Case Study 2—Using the Noise and Health Scientific Evidence Base to Inform Recommendations for National Noise Limits for Road, Railway and Aircraft Noise in Switzerland, by Mark Brink

In Switzerland the basic principles of the national noise abatement regime were established by the Environment Protection Act (1985) and the Noise Abatement Ordinance (1987). These contain regulations on precautions against noise, requirements for noise protection in new installations and for the renovation of existing installations, as well as requirements for construction zones and buildings with noise-sensitive rooms in noise-polluted areas. The fundamental principle is that the population should be protected from harmful or annoying noise. What is considered to be harmful or annoying must be defined by the government in the form of exposure limit values. To inform such values, a permanent expert commission (Federal Noise Abatement Commission) is in place that observes the scientific evidence base in the field of noise epidemiology.

The commission's last comprehensive report with updated recommendations for transportation noise exposure limits (for road, rail, and aircraft noise) was issued end of 2021 (https://www.eklb.admin.ch/de/documentation/berichte). However, preliminary work started back in 2007, where an interdisciplinary project team was commissioned to carry out a comprehensive study to clarify how the various relevant elements of noise abatement in the country (technology, operations, acoustics, health effects and regulations) have developed since the limit values came into force (1987) and whether there was a justified need for a detailed review and possibly an update of these limits. This preliminary work served as the basis for the launch of the multidisciplinary SiRENE study (2014–2020), which investigated the effects of noise exposure from road, rail and air traffic on annoyance, sleep and cardiometabolic morbidity and mortality in Switzerland. At roughly the same time the WHO Environmental Noise Guidelines for the European Region (World Health Organization Regional Office for Europe 2018) and particularly, their so called "evidence reviews" (Lercher et al. 2018), have been developed and published. With the SiRENE study, the WHO evidence reviews and other more recent international studies, the basis was laid on which the commission could examine whether the exposure limits for road, rail and aircraft noise still meet the legal requirements or whether and to what extent these limits need to be adjusted.

For the derivation of new limit values, the Noise Abatement Commission adopted a similar, but slightly different, heuristic as did the WHO for informing their guideline values. That means, basically, setting acceptable risks based on disability weights and identifying corresponding exposure limits on exposure–response relationships. The limit values were set at a level at which there was a "scientifically well substantiated increased risk of adverse health or annoying effects compared to a sufficiently low, 'uncritical', level". The commission explicitly considered two categories of effects that were both given the same weight for deriving the limit values: self-reported ("subjective") effects like annoyance and self-reported sleep disturbances on one hand, and on the other hand ("objective") cardiometabolic effects for which the evidence was considered scientifically sound enough, namely ischemic heart disease (IHD), diabetes, and cardiovascular disease mortality.

The Swiss law implies that a limit value only satisfies the legal requirements if it limits both harmful (i.e. "objective") and annoying (i.e. "subjective") noise effects. In other words, if an annoying noise sets in before becoming harmful, the limit value must be based on that threshold and vice versa. Therefore the two categories were in a first step treated separately and two separate limit values for each noise source (road, rail, air) and time period (day, night) were determined. Then, for each noise source and time period the respective lower of the two values (either from the "objective" or the "subjective" category) was adopted as the final limit value.

The recommended new limit values were not yet in force at the time of publication of this book, and the political process of their introduction (or rejection) is expected to take several years.

8.5   Health Risk Assessment as a Tool to Support Sound Governance

The main purpose of a health risk assessment (HRA) is to estimate and communicate the potential adverse health effects resulting from human exposure to a particular hazard—in this case, environmental noise (World Health Organization Regional Office for Europe 2018). One of the earlier examples of a noise HRA was the Burden of Disease from environmental noise (World Health Organization Regional Office for Europe and JRC 2011), published jointly by the WHO and the European Commission Joint Research Centre. This report expressed the burden of ill health due to noise in terms of Disability Adjusted Life Years (DALYs), a term that is widely used in public health and addressed extensively in Chaps. 4 and 5. The European Environment Agency has subsequently published a comprehensive HRA for European countries which uses a similar methodology but with updated evidence (European Environment Agency 2020).

Noise HRAs are increasingly being carried out at a national or regional level. The ICBEN 2023 review on policy and economics (Fenech and Janssen 2023) identified 13 assessments published in the scientific literature carried out in China, Ecuador, Estonia, Finland, Germany, Ghana, Iran, Pakistan, Spain, and Russia. Methodological guidance for carrying out noise HRAs has been published by several organizations, including the National Institute for Public Health and the Environment (Van Kamp et al. 2018), the European Environment Agency (2020) and Health Canada (2017). Developing useful health indicators for these HRAs can be a challenge—see the text box below.

The Challenge of Developing Useful Environmental Health Indicators to Inform Sound Governance
From WHO (Geneva 1999) Environmental Health Indicators—Framework and Methodologies (World Health Organization and Briggs 1999)
    Environmental health indicators are useful to help agencies and practitioners support and monitor policy on environment and health at all levels—from the local to the international. Indicators can:

- monitor trends in the state of the environment, in order to identify potential risks to health;
- monitor trends in health, resulting from exposures to environmental risk factors, in order to guide policy;
- compare areas or countries in terms of their environmental health status, so as to help target action where it is most needed or to help allocate resources;
- monitor and assess the effects of policies or other interventions on environmental health;
- help raise awareness about environmental health issues across different stakeholder groups (including policy-makers, health practitioners, industry, the public, the media);

- help investigate potential links between environment and health (e.g. as part of epidemiological studies), as a basis for informing health interventions and policy.

To be effective, indicators must satisfy a number of different criteria. They must provide a relevant and meaningful summary of the conditions of interest to their users. They must be transparent, testable and scientifically sound. If they are to detect variation or change, they must be sensitive to real changes in the conditions they measure, yet robust enough not to be overcome by uncertainties in the data used. They must also be cost-effective to compile and apply.

Many of these criteria are, to some extent, mutually incompatible, which is why indicators are difficult to design. The ultimate need for cost-effectiveness, for example, often means that indicators must be developed on the basis of data which already exist, however such data have often been collected for specific purposes, and are therefore not ideal for other applications. The need for clarity and ease of understanding also implies that indicators must often condense large amounts of data into a brief, simple and unambiguous message. The need for scientific validity, on the other hand, requires that this process must not go too far: indicators must simplify without distorting the underlying truth, and without losing the vital connections and interdependencies which govern the real world.

8.6 Towards More Equitable Sound Governance

Governance frameworks and mechanisms related to sound have, to date, rarely taken social variations of environmental quality and health into account. Such variations are influenced by environmental and planning institutions, but are often beyond citizens' control. Control over one's environment is a vital functional capability, which encompasses political participation, (property) rights and power to influence the environment. If environmental (noise) conditions infringe individuals' control over their environment, the resulting health inequalities can be considered unfair and unjust.

For a sound governance approach that leads to a trust-generating environment, Riedel et al. (2021) propose three fields of action:

(1) developing multi-/interdisciplinary training for students and practitioners that aligns the "traditional" noise impact/noise control approach with the salutogenic approach (see also Fig. 8.1), and helps strengthen the evidence base on the interlinkages of acoustic and non-acoustic factors from a local health equity and environmental justice perspective (including consideration of vulnerable populations);
(2) introducing comprehensible information and inclusive participation methods;
(3) creating supportive institutional frameworks and governance modes (see for example the concept of proportionate universalism (Carey et al. 2015)).

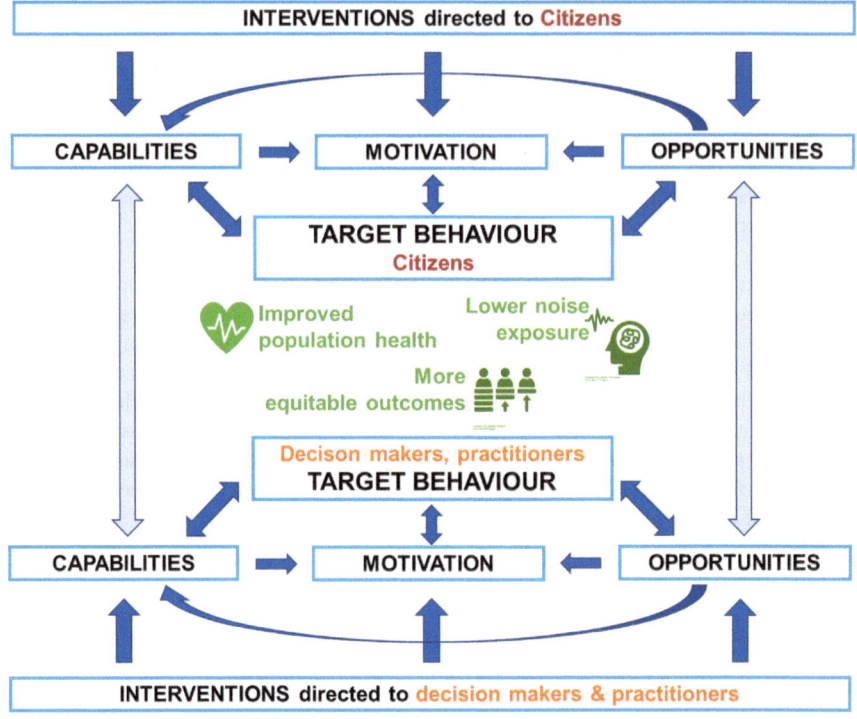

Fig. 8.3 A conceptual framework based on the COM-B model (Michie et al. 2011) for sound governance interventions targeting citizens and decision-makers. Equity icon by Adrien Coquet, noise icon by Kamin Ginkaew (both from the Noun Project, licensed under CC-BY 2.0)

Within this context, the Behaviour Change Wheel (BCW) (Michie et al. 2011) can be used as a strategic map to encourage more inclusive participation by inducing changes in behaviour of all involved in sound governance (see Fig. 8.3).

Involvement and behaviour change is desirable in all groups of society that produce, manage or are exposed to or affected by noise. Main target groups of interventions are citizens, decision-makers and practitioners, and enterprises that generate noise through their business (either directly and indirectly). Such interventions would need to take place in parallel with broader discussions on how to make economic activities more sustainable. In the following section, we illustrate the Behaviour Change Wheel by focusing on citizens and decision-makers/practitioners. Behaviours that interventions could target are, for example:

- for citizens: reducing personal exposure in daily life, engaging in consultations for transport infrastructure projects and reducing the use of personal motorised vehicles;
- for decision-makers/practitioners: establishing inclusionary and transparent planning procedures, applying proportionate universalism to noise policies, improving public transport, cycling routes and safety.

The behaviours in these two target groups are closely related. For example, citizens may find it difficult to reduce their use of motorised vehicles without suitable alternatives, and decision-makers may prioritise motorway expansion in response to an increasing demand for motorised traffic.

The BCW offers a systematic approach for planning behaviour change interventions in an integrated, context-sensitive way. It is a framework designed to aid intervention designers to identify intervention functions and policy categories that can bring about change by understanding factors that influence behaviour. A behavioural model describing behavioural conditions is therefore at the core of the BCW: the **COM-B system** (Michie et al. 2011). It presents three main components that condition a behaviour. The components are:

- **Capability** ("the individual's psychological and physical capacity to engage in the activity concerned");
- **Opportunity** ("all the factors that lie outside the individual that make the behaviour possible or prompt it"); and
- **Motivation** ("all those brain processes that energize and direct behaviour").

All three components should be targeted to achieve environmental justice and health equity.

Capability

This component includes psychological and physical capabilities. Physical capabilities are particularly relevant when it comes to mobility decisions (routes and transport modes) or physical efforts needed to express oneself successfully in participation formats offered by the municipality, for example. Psychological capabilities can be seen as perceived behavioural control,[1] meaning that an individual perceives the target behaviour as within her/his reach ("behavioural controllability") and is self-confident to perform this behaviour ("self-efficacy") (Ajzen 2002). Examples include psychological capacities that highlight knowledge, skills and self-efficacy that citizens and decision-makers/practitioners need to engage in the target behaviour, as inspired by Köckler (2017) and the Guidelines on Digital Inclusion of the Eurocities' CitiMeasure Project (https://citimeasure.eu/) on "Using citizen measurements to create smart, sustainable, and inclusive cities". An understanding of how to interpret and apply scientific findings and national legislation to local settings is equally relevant for both citizens and decision-makers/practitioners. However, psychological capabilities are role-specific. For example, vulnerable population groups may be reluctant to take part in political discourse, whereas decision-makers and practitioners may be reluctant to gain first-hand experience of citizens' capabilities, opportunities and motivations.

---

[1] Perceived behavioural control is originally presented as "reflective" motivation (see below) in the BCW framework. However here we frame it as psychological capability, given the conceptual link between knowledge, skills and self-efficacy in the Theory of Planned Behaviour as applied by Köckler (2017) in the environmental justice context.

Motivation

The motivation component includes reflective and automated processes. An example of reflective processes is the commitment to, or belief in the value of adopting target behaviours and the expected reward. For example, citizens try to anticipate whether a target behaviour such as campaigning against a specific noise source is worth a try. Outcome expectancy can relate to the feedback from relevant others, but also to the enjoyment when engaging in the behaviour.[2] Decision-makers and practitioners may have outcome expectancies for changing one-directional consultation into transparent and inclusive procedures.

Automated processes can originate from habitual patterns, emotional responses and impulses, as well as unconscious triggers. These stem from "associative learning and/or innate dispositions" and are closely linked to individual personal attributes.[3]

Opportunity

The opportunity component of the behavioural model is the context. Behaviour can only be understood in relation to context, and this is the starting point of intervention design (Michie et al. 2011). Structural, institutional and social opportunities influence citizens' psychological capabilities and motivation directly or indirectly by fostering or impeding behavioural practice. For example, social opportunities from citizens' *cultural milieu* shape their (reflective) motivation by conveying attitudes towards democracy, citizenship, authorities and environmental concerns, social norms and cultural values.

Regarding decision-makers/practitioners opportunities are linked to planning culture. On the one hand, opportunities may be specified by formal contexts, such as binding environmental (noise) standards and rules of resource allocation. On the other hand, informal contexts as expressed by an attitude towards scientific evidence, the professional development in the field of noise, health, non-acoustic factors, recognition of intersectionality,[4] participation and inter-sectoral practice, and collaborative working. These opportunities influence decision-makers' and practitioners' motivation, capabilities, and finally the performance of the target behaviour.

Capabilities, Motivation and Opportunities in Combination

The capability and opportunity components can promote or suppress the motivation component as well as a target behaviour directly. Practising the target behaviour has an impact on the three components, i.e. it can reinforce or inhibit the components. By

---

[2] Within the frame of the Theory of Planned Behaviour, reflective processes could be attributed to attitudes towards the behaviour and subjective norm.

[3] Reflective processes are, in our understanding, connected to what has been termed as "stimulus expectancies" in the Cognitive Activation Theory of Stress by Ursin and Eriksen (2004). These types of expectancies are more related to coping strategies of cognitive distortion and resignation as opposed to active coping strategies that are needed to enact behavioural change.

[4] Intersectionality is a theoretical framework that illustrates how intersecting social identities at the individual level reflect structural patterns of privilege and oppression along different axes, such as race, gender, social class, sexuality, and ability (Williams et al. 2023).

way of an example, previous experience of performing the behaviour and its perceived effects will influence reflective processes associated with outcome expectancies.

The capabilities, opportunities, motivation and resulting behavioural change(s) of target groups are likely to also interact with each other. For instance, citizens starting civic engagement against a noise source may alter decision-makers'/practitioners' reflective processes that favor inclusionary and transparent planning procedures. This might be accompanied or fostered by an increase in skills in the field of citizen science on both sides. The CitiMeasure Guidelines (https://citimeasure.eu/) structure their recommendation for digital inclusion and behavioural change by target groups, i.e. cities, citizen science initiators or facilitators, information professionals, and citizen scientists. To achieve this mutual enrichment between citizens and decision-makers/ practitioners, continuity of "co-creative" behavioural practice is needed—instead of fragmented, occasion-driven events (Bell and Carrick 2018). Table 8.1 provides some example considerations for both citizens and decision-makers.

8.6.1 COM-B and Sound Governance

An intervention to stimulate behavioural change can target one or more components of the COM-B model. In the Behaviour Change Wheel (BCW) model, nine intervention functions are proposed, together with seven policy categories that can support the delivery of these intervention functions. Table 8.2 shows practical examples of the BCW intervention functions applied to decision-makers to help them establish inclusionary and transparent planning procedures (target behaviour change).

8.7 Concluding Remarks

Going through the myriad of regulatory processes, mechanisms and organizations presented in this chapter, one can easily feel disheartened by the complexity of the system and the difficulty in understanding where and how to influence environmental actions and outcomes. The WHO 2018 noise guidelines (World Health Organization Regional Office for Europe 2018) are very clear that: *"Preventing noise and related health impacts relies on effective action across different sectors: health, environment, transport, urban planning and so on"*. But ultimately all these different sectors have the same thing in common—they are all aiming to achieve *"peace and prosperity for people and the planet, now and into the future"* (https://sdgs.un.org/goals). A sound approach to noise and health offers a great opportunity to help us get there.

Table 8.1 Elements of capability, motivation and opportunity in relation to behaviour change for citizens and decision-makers for more equitable sound governance

COM-B (Michie et al. 2011) domain	Examples for citizens	Examples for decision-makers/ practitioners
Capability	• Confidence to participate in the political discourse or to try alternatives to daily routines • Skills in the context of citizen science: data science, management, co-creation and transferability • Understanding of what is at stake: scientific evidence, political reasoning, principles and procedures	Skills and knowledge to: • Understand scientific evidence • Analyse and evaluate planning aspects from an intersectionality perspective (e.g. (Williams et al. 2023)) • Translate planning principles into action • Monitor distributional effects of interventions • Develop a co-creative participation design
Motivation	• Attitude towards the target behaviour(s)/outcome (influenced by previous participation experience, trust) • Expected feedback from relevant others • Match between role/self-identity and target behaviour • Enjoyment when engaging in the behaviour	• Commitment to reaching the highest environmental standard possible for all • Expected reward from relevant colleagues and superiors, higher planning levels • Role and identity as planners/decision-makers
Opportunity	• Working patterns and caring responsibilities • Financial means • Accessibility of infrastructure, public services to support target behaviours • Participation opportunities in real-world settings • Support from relevant others to engage in target behaviour	• Environmental (noise) standards and planning principles (net gain, precautionary, health equity, etc.) • Allocation of resources (personnel, finances) • Attitude towards science among leaders/in the occupational context • Expertise in the field of noise, health, non-acoustic factors, etc. in the occupational context • Recognition of intersectionality • Inter-sectoral working practice

Table 8.2 Example interventions to help decision-makers establish inclusionary and transparent planning procedures

BCW Intervention function	Example actions targeting decision-makers and practitioners
Education—increasing knowledge and understanding	• Explaining strengths and limitations of exposure–response relations, • Presenting non-acoustic factors as a means to contextualise exposure–response relations, • Recognising different types of scientific evidence, whilst appreciating residents' contributions, incl. citizen science initiatives, as a complementary type of 'local evidence', • Understanding structural drivers of unequal distributions of noise exposures and (potentially compensating) environmental resources, • Introducing the concept of intersectionality with focus on how to recognise, avoid and alleviate discrimination based on personal characteristics • Structural reasons for unequal coping opportunities among residents
Modelling	• Presenting "best practice" examples from similar cities
Training–imparting skills	• Skills on how to apply the net gain principle, vulnerability of the population principle (Köckler 2017), precautionary principle, and polluter pay principle to achieve more health equity in planning process
Enablement	• Enrol coaches/mentors to assist practitioners in situations where application/realization of equitable planning principles is difficult • Reflect on what an "intersectionality approach" could reveal in a specific planning-related situation
Incentivisation	• Participate in prestigious competitions or apply for relevant certifications (e.g. comparable to the DGNB system for "sustainable districts" (https://www.dgnb.de/en/certification/districts) • Award additional budgets if the municipality engages in target behaviours (e.g. transparent participation procedures)
Coercion	• Link departmental budgets to performance indicators linked to planning procedures

(continued)

Table 8.2 (continued)

BCW Intervention function	Example actions targeting decision-makers and practitioners
Persuasion	• Demonstrating where (noise action) planning can benefit from active participation or citizen science and where intersectoral collaboration can lead to more funding opportunities
Environmental restructuring	• Introducing formats for inter-sectional exchange to inform planning decisions • Develop planning decision-making systems dedicated to equity and intersectionality concerns, incl. transparency in reasoning (e.g. explaining how equity issues have been dealt with)

Acknowledgements The authors would like to thank Lisa Woodland Ph.D. for her advice on the application of the Behaviour Change Wheel. We are grateful to colleagues, fellow authors, editors and reviewers who have helped shape a better chapter through careful and constructive feedback and stimulating comments. Special thanks go to Fred Woudenberg, Irene van Kamp and Marion Burgess.

References

Ajzen I (2002) J Appl Soc Pyschol 2002(32):665–683
Barton H, Grant M (2006) J R Soc Promot Health 126(6):252–253. ISSN 1466-4240
Bell D, Carrick J (2018) In: Holifield R et al (eds) The Routledge handbook of environmental justice, 1st edn. Routledge, London, UK; New York, USA, pp 101–112
Brown and Van Kamp (2017) Int J Environ Res Public Health 14(8):873. https://doi.org/10.3390/ijerph14080873
Carey G et al. (2015) Int J Equity Health 14:81. https://doi.org/10.1186/s12939-015-0207-6
Chauhan R, Shrestha A, Khanal D (2021) Noise pollution and effectiveness of policy interventions for its control in Kathmandu, Nepal. Environ Sci Pollut Res 28:35678–35689. https://doi.org/10.1007/s11356-021-13236-7
Choi E et al. (2022) Arch Environ Occup Health 77(2):149–160
Clark C et al. (2021) J Acoust Soc Am 150:3362–3373
Commission to the European Parliament and the Council (2011) https://eur-lex.europa.eu/legal-content/EN/TXT/?uri=CELEX:52011DC0321
Corvalán C et al. (1996) Geneva: UNEP, USEPA and WHO, pp 19–53. https://apps.who.int/iris/handle/10665/62988
COST Action TU0901 https://www.costtu0901.eu/index.html
Dahlgren G, Whitehead M (2007) WHO Regional Office for Europe. http://www.euro.who.int/_data/assets/pdf_file/0018/103824/E89384.pdf
Department for Environment Food and Rural Affairs (2014) National Noise Attitude Survey 2012—NO0237. https://randd.defra.gov.uk/ProjectDetails?ProjectID=18288
Directive 2002/49/EC of the European Parliament and of the Council (2002) https://eur-lex.europa.eu/legal-content/EN/TXT/?uri=CELEX:32002L0049

Eidgenössische Kommission für Lärmbekämpfung. https://www.eklb.admin.ch/de/documentation/berichte

European Commission (2021a) https://op.europa.eu/en/publication-detail/-/publication/f4cd7465-a95d-11eb-9585-01aa75ed71a1

European Commission (2021b) https://eur-lex.europa.eu/legal-content/EN/TXT/?uri=CELEX:520 21DC0400

European Commission (2016) Directorate-General for Environment, Evaluation of directive 2002/49/EC relating to the assessment and management of environmental noise: final report. Publications Office. https://doi.org/10.2779/171432

European Commission, J.R.C., Zero Pollution Outlook (2022) EUR 31248 EN, Publications Office of the European Union, Luxembourg. ISBN 978-92-76-57575-7, https://doi.org/10.2760/778012, JRC129655

European Environment Agency (2020) EEA Report No 22/2019. https://www.eea.europa.eu/publications/environmental-noise-in-europe

European Environment Agency (2020) https://www.eea.europa.eu/publications/health-risks-caused-by-environmental

Fenech B, Janssen S (2023) In: Proceedings to 14th ICBEN congress on noise as a public health problem

Fenech B et al. (2021) Proceedings to 13th ICBEN congress on noise as a public health problem

Flacke J et al. (2022) Front Sustain Cities 4:835534

Ghimire N et al. (2019) Asian J Med Sci 10(5):49–54

GoN (2019) The Environment Protection Act, 2019. Government of Nepal

Hauptvogel D et al. (2021) Int J Environ Res Public Health 18(14):7399. https://doi.org/10.3390/ijerph18147399

Health Canada (2017) H129-54/3-2017E-PDF. https://iaac-aeic.gc.ca/050/documents/p80054/119378E.pdf

http://www.icben.org/Proceedings.html

https://citimeasure.eu/

https://sdgs.un.org/goals

https://unece.org/press/new-un-regulation-keeps-silent-cars-becoming-dangerous-cars

https://unece.org/press/unece-world-forum-harmonization-vehicle-regulations-tightens-vehicles-noise-limits-and-adopts

https://unece.org/sustainable-development/press/new-un-regulation-harmonizes-reverse-warning-sound-vehicles

https://www.dgnb.de/en/certification/districts

https://www.gov.wales/new-powers-tackle-air-and-noise-pollution-will-lead-cleaner-healthier-and-greener-future

https://www.iso.org/standard/52161.html

https://www.iso.org/standard/69864.html

https://www.iso.org/standard/74048.html

https://www.iso.org/standard/75267.html

https://www.iso.org/standard/77742.html

https://www.iso.org/standard/84809.html

ICAO. https://www.icao.int/environmental-protection/pages/noise.aspx

Jephcote C, Clark S et al. (2023). https://doi.org/10.1016/j.envint.2023.107966

Joshi S et al. (2003) Kathmandu Univ Med J (KUMJ) 1(3):177–183

Köckler H (2017) Umweltbezogene Gerechtigkeit. Anforderungen an eine zukunftsweisende Stadtplanung. Peter Lang GmbH, Frankfurt (Main), Germany

Kropp et al. (ed) (2016) Chalmers University of Technology, Division of Applied Acoustics. https://publications.lib.chalmers.se/records/fulltext/242257/local_242257.pdf

Laplace J et al. (2022) Canadian Public Policy 2022 vol 48:1, pp 74–90

Lemos MC, Agrawal A (2006) Annu Rev Environ Resour 31(1):297–325. https://www.annualreviews.org/doi/10.1146/annurev.energy.31.042605.135621

Lercher et al. (2018) Special issue "WHO Noise and Health Evidence Reviews". Int J Environ Res Public Health

Leylekian L et al. (ed) (2022) https://doi.org/10.1007/978-3-030-91194-2

Malla L (2017) The Kathmandu Post XXV

Manandhar M et al. (1987) A reported submitted to Nepal National Committee for the Man and Biosphere, Kathmandu

Michie S et al. (2011) Implement Sci 6:42. https://doi.org/10.1186/1748-5908-6-42

National Trust. Octavia Hill: her life and legacy https://www.nationaltrust.org.uk/discover/history/people/octavia-hill-her-life-and-legacy

Nepal Gazette (2012) National Sound Quality Standard. Government of Nepal

Nepal Health Research Council (2003) https://elibrary.nhrc.gov.np/handle/20.500.14356/151

Niemann H, Maschke C (2004) LARES final report noise effects and morbidity. World Health Organization

OECD (2022) https://legalinstruments.oecd.org/public/doc/35/35.en.pdf

Peeters B, Nusselder R (2019) European network of the heads of environment protection agencies (EPA Network) M+P.BAFU.18.01.1

Peris E, Fenech B (2020) Sci Tot Environ 748. https://doi.org/10.1016/j.scitotenv.2020.141040

Perna M et al. (2021) Int J Environ Res Public Health 2022 19(1):173. https://doi.org/10.3390/ijerph19010173

Rasmussen B, Machimbarrena M (2019) Proceedings to InterNoise, pp 7751–7762(12)

Riedel N et al. (2021) Transp Res Interdiscip Perspect 11:100445. https://doi.org/10.1016/j.trip.2021.100445

Riedel N et al. (2022) Cities Health 6(2):258–266

Robinson T et al. (2015) Int J Occup Environ Health 21(1):14–22

Sanju HK, Kumar P (2016) Indian J Otol 22(3):162–167

Schreckenberg D et al. (2018) Int J Environ Res Public Health 15(5):971. https://doi.org/10.3390/ijerph15050971

Shrestha C, Shrestha B (1985) A reported submitted to National Committee for Man and Biosphere. Nepal, Kathmandu

Shrestha I et al. (2011) Kathmandu Univ Med J (KUMJ) 9(36):274–278

Shrestha R et al. (2022) Int J Environ Res Public Health 19(8):4752

UN Environment Programme (2022) https://www.unep.org/resources/frontiers-2022-noise-blazes-and-mismatches

United Nations Environment Programme and World Health Organization (1980) https://wedocs.unep.org/20.500.11822/29536

Ursin H, Eriksen HR (2004) Psychoneuroendocrinology 29(5):567–592. https://doi.org/10.1016/S0306-4530(03)00091-X

Van Kamp et al. (2018) National Institute for Public Health and the Environment (Netherlands). https://www.rivm.nl/bibliotheek/rapporten/2018-0121.pdf

Williams PC et al. (2023) J Am Plan Assoc 89(2):167–174. https://doi.org/10.1080/01944363.2022.2079550

World Health Organization (1999) https://www.who.int/publications/i/item/a68672

World Health Organization (2021a) https://www.who.int/publications/i/item/9789240020481

World Health Organization (2021b) https://www.who.int/activities/making-listening-safe

World Health Organization (2022) https://www.who.int/publications/i/item/9789240043114

World Health Organization and Briggs DJ (1999) https://www.who.int/publications/i/item/WHO-SDE-OEH-99.10

World Health Organization and International Telecommunication Union (2019) https://www.who.int/publications/i/item/safe-listening-devices-and-systems-a-who-itu-standard

World Health Organization Regional Office for Europe (2009) Copenhagen https://www.euro.who.int/__data/assets/pdf_file/0017/43316/E92845.pdf

World Health Organization Regional Office for Europe and JRC (2011) https://www.who.int/publications/i/item/9789289002295

World Health Organization Regional Office for Europe (2018) https://apps.who.int/iris/handle/10665/279952
Yokoyama S, Kobayashi T (2022) Proceedings to InterNoise 2022

Chapter 9
Sonic Ecology and the Role of Sound Art in Creating a Healthy Urban Environment

Marcel Cobussen

Tuesday, December 2022

I'm staring out the window. It is still a bit dark. Through one window I see a few clouds and a sky that promises to become blue soon. The other window, however, offers a view colored in several shades of grey. In the foreground, bare tree crowns and family homes, many of their windows still framing closed curtains.

I'm pondering the start of this essay. I look at the instructions. A footnote tells me that I'm allowed to add up to five illustrations. The remark raises vaguely formulated questions in my mind, something like "When will illustrations ever represent sounds better than the sounds themselves?" and "Why didn't they offer the possibility to include audio files?" This, in turn, makes me think of the first sentences I wrote: no mention of what I'm hearing either. Outside, the cooing of a pigeon, the almost-human screams of a few seagulls, and the wind blowing around the house. Inside, the buzzing of the computer, fingers ticking on the keyboard, my clothes rubbing against my desk … and the sound of my study—hard to describe, as it is a sound bordering on silence, almost inaudible but present nevertheless. Like the neighborhood, my family members are still sleeping, so all in all it is very quiet.

My mind wanders to you, the reader, whose eyes are currently registering these words. What do you hear at this very moment? Would you be able to hear your eyes moving? Are you nestled in a comfortable armchair with this book while your favorite music is playing? Bach? Chet Baker? Arctic Monkeys? Or are you traveling, reading while the plane you're in takes off, causing quite some noise to which you will get accustomed in a few minutes? Or not! Of course, the possible situations in which you read what I wrote back then, in my study on the top floor of my house on that chilly morning in December, are endless. But were you already aware of

M. Cobussen (✉)
Faculty of Humanities Academy of Creative and Performing Arts, Leiden University, Leiden, The Netherlands
e-mail: MA.Cobussen@hum.leidenuniv.nl

© The Author(s) 2025
I. van Kamp and F. Woudenberg (eds.), *A Sound Approach to Noise and Health*, Springer-AAS Acoustics Series, https://doi.org/10.1007/978-981-97-6121-0_9

your sonic environment? What did you hear before I drew your attention to the sounds that surround you now? And do those sounds affect your reading? How? Does Bach encourage you to read more slowly and concentratedly? Do the sounds of the airplane prevent you from clearly focusing on these words? Or is it the other way around: the beautiful music is distracting, while you are able to filter out the less pleasant sounds, even when they are loud? I don't know, but I will come back to these questions further on in this essay ...

9.1   Sound and Well-Being

I have just reread the opening chapter of this book. Irene and Fred are right: living in a big city is not always pleasant for the ear. Sirens; the thumping sounds of construction work; the almost constant buzz of traffic sounds, frequently interrupted by the noise of airplane engines, the occasional screaming and yelling of nighttime clubbers and partygoers, etc. But these are just the obvious loud sounds that can disturb one's mood, activities, or even health. Less noticeable sounds can have a comparable impact on one's life. A friend of mine once told me how the ticking of a cord against a flag pole kept him awake at night. High-pitched squeaking is another example of low-quality sounds, whether they are made by humans, animals, tram and train tracks, or utensils. Even the calling songs of an indigenous Cuban cricket seemed to induce nausea and headaches, as American diplomates in Havana experienced in 2016 (Zimmer 2019). Even if they are relatively low on the decibel scale, random, non-continuous sounds prove to be quite alarming to our nervous systems as well. And recently there has been an increase in awareness that even barely audible sounds, or sounds that cannot be captured by the human ear, can negatively impact people's health. Low-frequency sounds or the vibrations of electro-magnetic fields are claimed not only to cause pain but also disruption of the vestibular system's signals to the brain, which can result in vertigo, nausea, loss of balance, or fatigue (Goodman 2010, pp. 183–188; Johnson 2023, pp. 172–175). These low-frequency tones are known as *the Hum* and, although their source is often unclear, they are primarily reported from regions with the greatest development of electronic infrastructure (Epstein 2020, p. 208).

However, in their introduction to this book, Irene and Fred also make clear that sounds are not only the source of doom and gloom. Loud sounds, for example, can also be pleasant and actively sought out. For a long time, noisy industrial sounds connoted economic progress and increasing prosperity; concerts of, especially, pop and rock music, are often performed at high decibel levels in order to immerse the audience in a pool of sound, thereby enhancing their aesthetic experience; the loud chants of football supporters in a stadium are meant to intimidate the opposing team as well as to create a sense of belonging; and for centuries firecrackers have been set off to celebrate specific events as well as to chase away evil spirits.

In short, the relation between sound and health or well-being is ambivalent, to say the least: the same sound or decibel level can be pleasant for one person while being intolerable for another. It seems that experiencing sound is an individual, subjective affair. However, things are more complicated: the same person can experience a specific sound on a specific level as obnoxious in one context while enjoying it in another.[1] In other words, merely saying that the perception of sound is subjective is too limited; both the existence of sound and the experience of sound are relational, that is, depending on specific situations.[2]

Hence, building on some later paragraphs in Irene and Fred's opening chapter, my contribution will focus less on individual health issues and more on the influence sound has on social well-being and well-being in general. This implies that sound will be connected to social, economic, cultural, ecological, political, ethical, and aesthetic issues.[3] More theoretical and reflective passages will alternate with descriptions of concrete projects in which I was involved, projects that involved analyzing the sonic ambiance of public urban spaces and recommending interventions to improve this ambiance and/or the ways it is or can be perceived and experienced. Thus, the emphasis of this chapter will be on collective health and well-being as well as the positive role sound can play. The hypothesis I would like to defend here is that an important opportunity for improvement can be found in supporting more and more attentive interactions between human beings and their sonic environment.[4] The challenge for urban planners is to contribute to (re)creating sites of belonging and connecting, using, among other modes of sensory interventions, sound, so that both human and nonhuman bodies might be able to affect one another in multiple and versatile ways. In more general terms, urban planners should attempt to expand the affective potential of the (sonic) environment; and to be more concrete, a sonic environment of high quality can be defined as an environment that increases the capacity of both human and nonhuman agents to act and to relate. Expanding the affective potential of a site can be called a micro-political act, as it challenges those forces that restrict the possibilities to create spaces of interaction and experiential diversity (Lacey 2016, p. 41).

---

[1] As I wrote elsewhere, a listener always hears from a particular position: either outside or inside, in a familiar or an unfamiliar environment, attentively or distractedly, in a loud or quiet environment, etc., and this affects the evaluation of a sound or (sonic) ambiance as a whole (Cobussen 2022, pp. 18–19).

[2] You don't just hear the sounds of a car. You hear whether the car is passing by or standing still; whether it is being driven forward or backwards; whether it is a small or a large car; whether it is being driven through a narrow or a wide street, through a green environment or between high buildings, and of which materials these buildings are made; whether the car is traversing a dry or a wet street surface; whether the surface is made from asphalt, bricks, or semi-paving, and if there is just one car or many more (Cobussen 2022, p. 18).

[3] Coming back once more to the example of a car: becoming aware, through listening, of the number of cars, their brands, the street surface, and the overall environment provides extensive information about the social, economic, and cultural contexts in which the car sounds appear (Cobussen 2022, p. 19).

[4] Although the term "sonic" just seems to be the Latin equivalent of the Greek "acoustic", the (recent) history of sound studies shows that, in general, "acoustics" is often referring more to physics while "sonic" emphasizes the way sounds are perceived and understood by living beings. As the overall perspective of this chapter is more on the social aspects and influences of sounds, I have decided for the term "sonic environment."

However, as I will make clear toward the end of this essay, improving well-being by paying attention to the sonic environment is not only the responsibility of urban planners, project developers, or (local) officials. Residents, users of specific (semi-) public spaces, and (sound) artists are groups that can also be mobilized to transform sonically unpleasant sites in order to increase general, social, or political well-being.

9.2   The Sounding City

What is a city? How can a city be defined? Or, to restrict, delimit, and specify this question a bit more: What is a city in relation to sound? Even though this adjusted formulation is still too broad to be dealt with in this context, a few crucial remarks can be made here.

In the first place, a city is formed by both human and nonhuman agents: a city is an assemblage of buildings, roads, bridges, squares, parks, lakes or waterways, etc.; a city is simultaneously an assemblage of humans and other living beings (animals, trees, plants) who are acting within this built environment and, through their activities, contributing to the creation of that environment; a city is also an assemblage of social, political, economic, ethnic, aesthetic, and ethical encounters, exchanges, oppositions, or conflicts.[5] In that sense, a city is not a stable entity, reducible to its geographic coordinates, its built environment and existing infrastructure; a city is a vibrant organism with continuously changing interactions between stable and less stable elements, interactions which are, moreover, co-determined by weather conditions, times of the day, days of the week, and the concrete particularities of the specific agents that are active or passive at specific moments and places (Cobussen 2016). Cities are thus established through flows of emergent interconnections generated by the establishment of affective relations between heterogenic components. All these agents or components are capable of acting and are at the same time also acted upon. Functions, properties, and identities of agents that together constitute a city emerge from the flows between these components, and the flows themselves range between chaos and relative (and often temporary) points of stability; together, these create the city as a relational net.

Specifically in relation to the sonic here: the city is a constantly vibrating entity, a permanently changing *sounding* city as motorized, bicycle, and foot traffic, music, sirens, ringing bells, human voices, chirping birds, dripping rain on various surfaces, rustling trees, rippling water, construction works, public events (from street protests to festivals), creaking bridges, etc., create a continuous and unpredictable urban symphony in which the more stable elements codetermine the degree of reverb, echo, reflection, attack, and decay. An urban sonic environment thus not only establishes multiple links between physical spaces, human bodies, and sound but also reflects

---

[5] In this respect, sociologist Richard Sennett, in (Sennett 2018) *Building and Dwelling: Ethics for the City*, distinguishes between two aspects of the city, the *ville* and the *cité*. Whereas the *ville* refers to the built urban environment, the *cité* deals with the ways of life of its inhabitants.

socio-economic structures and living standards. It sonically binds together human and nonhuman bodies, materials, ideas, ideologies, technologies, and social stratifications (Gandy and Nilsen 2014, p. 56 and 148).[6]

Second, humans and other living beings can use the more stable agents in a city in many different ways. To a certain extent, the way a city is physically structured implicitly disciplines its inhabitants. For example, the introduction of sidewalks made a separation possible between motorized traffic and pedestrians; today, the public benches in parks all too often have an armrest in the middle so that homeless people cannot lie down for the night there; and large urban planters or flower pots prevent cars from entering pedestrian zones. To focus more explicitly on sounds: the sirens of a police car or firetruck insist that car users make space for quick passage, and music coming from shops and audible on the street is meant to draw inside potential customers. Sidewalks, benches, flower pots, and music may improve the appearance and livability of a city, but, simultaneously, restrict and regulate the range of motion of inhabitants as well as the sonic atmosphere of a site.

However, people also occasionally find the liberty to ignore or subvert this disciplining function of urban design. For example, when wearing headphones and listening to their favorite music, they might behave (slightly) differently: ignoring traffic signs, singing along, adapting their walking speed to the music's rhythm, snapping their fingers. As urban sociologist Jean-Paul Thibaud writes with regard to the urban wanderer wearing a portable audio device:

A kind of tuning in is created between his ear and his step. New sonic territories are composed in the course of this mobile listening experience. As the body moves in synch with the music, the listener transforms the public scene and provides a new tonality to the city street. His footsteps seem to say what his ears may be trying to hide (Thibaud 2003, p. 329).

And historian Heike Weber in Gandy and Nilsen (2014, pp. 157–160):

The conjunction of urban masses with shrinking portable devices, and even wearable technologies, resulted in changing patterns of how to move around urban space and encounter others [...]. [This is in a way] a domestication of the acoustic environment, commonly defined as shared and public [...] [by combining] previously separated activities such as learning languages while shopping, listening to music while changing subways.

By turning the volume up or down, the wanderer-listener filters and enhances the events that take place on an urban site, events that give these sites their meaning. They navigate through more than one world at once—the one in which they listen and the one in which they walk—at the same time creating a disjunction between the visible and the audible. Social and cultural codes regarding how to behave in a public urban space are challenged, stretched, and redefined. While I work on this text, football supporters take possession of several squares in my hometown to celebrate the victory

---

[6] Here I use the concept of the body in a broad sense, encompassing the human body, the body of a building, the materiality of objects, but also the immaterial structures that compose a site. In that sense, this means there is also a social body, a cultural body, a political body, and an infrastructural body. Rather than restricting well-being in relation to sound to (the experiences of) individual human bodies and their perceptions of the soundscape, the sonic ecology I am proposing here considers the relationships between *all* bodies.

of their favorite team during the FIFA World Cup tournament, thereby also sonically occupying and appropriating these spaces with songs, car horns, and firecrackers. Simultaneously, elsewhere in the world, people take to the streets, producing sounds which are not usually heard there, giving voice to their discontent with the existing social order or political developments, and marking what is outside the boundaries of consensual political discourse, thereby turning noisy sounds into a constitutive act of political expression. All these examples indicate a certain interruption of the more or less normal, everyday relations between sounds and the city.

Whether positively or negatively, disrupting or disciplining, natural or artificial, loud or soft, sounds form an inextricable part of a city's ambiance and the lives of its inhabitants. Whether we like it or not, we are always surrounded by sounds, always immersed in a sonic environment. And, just as the air we breathe should be clean enough that it does not cause death due to air pollution, we also are in need of—and are actually entitled to—a healthy sonic atmosphere.[7] In that sense it is still striking that, while social conflicts over sounds are growing, in almost all stages of designing urban public sites, sound design is hardly given any attention apart from guaranteeing compliance with certain legal limits regarding noise pollution[8]; most often, sound in public environments is either regarded as a fait accompli—an inevitable side effect of certain activities or events—or as a problem. Only rarely is sound regarded as an integral and impactful element of an urban environment, an element that can be composed, directed, and adjusted, thereby also regulating how it affects the experience of people being exposed to it.[9]

Friday, October 2022

I'm standing on a floating pontoon in Rotterdam (the Netherlands), close to the southern bank of the river Maas. The pontoon lies in a small bifurcation of the Maas and is surrounded by medium–high buildings, mostly with the smooth façades that contemporary architecture still dictates. More apartment buildings will be erected in the years to come; densification is one, if not the only, solution to meet the demand for extra housing while sparing the few rural areas left around Rotterdam.

The pontoon, in its current (but still provisional) form of around 80 × 30 m, has a green area with grass, plants, and a few young trees, covering approximately 40%

---

[7] Once more, "healthy" in this context should not merely be understood as pertaining to an individual's mental and physical condition but should also encompass social, political, ethical, ecological, and aesthetic issues.

[8] "Acoustic gentrification" is an important factor in social conflicts over noise pollution. Historically, but also today, control over sound and silence have been used as forces of political class struggle. Generally speaking, only affluent residents can, more or less, ensure that their direct environment is selectively quiet. This "more or less" is a crucial addition here as project developers can perhaps install gates and cameras to separate the "haves" from the "have-nots" but can hardly prevent noise from penetrating those gates (Gandy and Nilsen 2014, pp. 206–208).

[9] As sonic theorist Gascia Ouzounian writes (in Gandy and Nilsen 2014, p. 166): "Developing a positive approach towards acoustic ecology is an important idea, which diverges from many historical soundscape projects that document noise pollution and other undesirable aspects of the acoustic environment".

of the pontoon's surface. The remaining surface consists of flat, hard, and smooth polished asphalt concrete with some wooden decks here and there. I clap my hands, an activity resulting in some impressive echoes; it is clear that the height of the surrounding buildings, the smooth surfaces, and the calm water create abounding reverberation. And it is equally clear that—for a place designed for social interaction and recreation—this echoing is less ideal. The function of the pontoon, as postulated and imagined by the urban planners, could very well be disrupted due to the rather low sonic quality: the hollow and superimposed sounds counteract the relatively relaxed atmosphere that the pontoon was (visually) designed to emit.

What can be done, sonically, to improve this space? How can this site be sonically designed so that the pontoon becomes a site for relaxation, for reading a book, for chatting with friends, or for putting your feet in the water on a hot day and (thereby) escaping the hecticness of the nearby city center and its main roads? Some first thoughts emerge: the relevant building façades could be covered by vegetation or detailing that serves to decrease the amount of reverberation and echo. Why does more than half of the pontoon's surface area consist of this hard, smooth surface? Why not use softer ground covers, such as grass or sand? More difficult to realize, perhaps, but eliminating parallel façades could reduce unnecessary resonances and standing waves, as would the introduction of volumes with round or irregular forms (Photo 9.1). And, would it be possible to mask unwanted sounds by adding more pleasant and non-disruptive ones, thereby shielding visitors from the dominant city sounds?

9.3 Redesigning the City Center of Rotterdam

A couple of years ago, Rotterdam designated a few locations that need to be redesigned within the coming decade in order for the city to become more climate neutral, ecologically solid, green, and therefore more livable, healthy, and pleasant. Redesigning these locations also explicitly meant paying attention to their sonic composition or ambiance. Even though most of the available time, energy, and money will still be invested in the visual design, the management teams of most of these projects did become convinced that without a decent sonic ambiance, their envisaged objectives would not be achieved.

Moreover, officials from the city developed the concept that—in addition to the well-established and implemented environmental noise management guidelines—policy regarding sound should also be based on the experiences and qualitative assessments of the sonic environment by its citizens. They managed to convince their political leaders that relying on decibel measurements only has its limitations in attempting to identify the negative and positive aspects of sounds in relation to a specific site. Besides these measurements and thorough analyses of the data acquired by these measurements, more insight needed to be gained regarding how people were *experiencing* these locations, whether they were satisfied with the current (sonic) situation or not, what their main grievances and preferences were, what should be

Photo 9.1 Floating pontoon at the Rijnhaven, Rotterdam. © Marcel Cobussen (2023)

preserved, etc. Once more—and putting it very simply—while sounds that don't qualify as (too) loud can arouse unpleasant reactions, some loud sounds may be experienced as characteristic of a specific site and should therefore be protected, for example, because of their historical and/or cultural value.

In order to gain more insight into how residents, tourists, and other users experience a particular public space in Rotterdam, these officials got in touch with me as chair of Auditory Culture at Leiden University with a special interest in the sounds of public urban spaces. I wrote a research proposal, created a small research team, and started working. On the basis of soundwalks (by the research team, with residents as well as urban planners, architects, and management) that were followed by semi-structured interviews; questionnaires (also online, on the website of Rotterdam); workshops in which various stakeholders could exchange ideas regarding their ideal sonic environment in relation to a specific site; additional measurements regarding the specific sonic characteristics of a site; and, of course, existing literature, several recommendations were formulated.

The point of departure for these researches was not (only) to identify which sounds were considered bothersome or intrusive. The main questions included: What do you hear? Which sounds are dominant and which ones are absent or hardly audible? How would you like this place to sound? Which sounds do you like? etc. In other words, sound was not presented or framed as a problem but as an opportunity, as an element within a cityscape that can—to a certain extent at least—be composed, designed, and taken care of. Answers to the questions formed the backbone of the advice that the researchers provided to the project managers. It is also important to note that the emphasis in the recommendations was not so much on quietness per se but, rather, on a better balance between tranquility and liveliness; after all, the designated locations are located in rather busy and noisy areas of the city. This also implied that already-existing urban soundscape design strategies would require some sort of recalibration to respond more effectively to the inevitable noises of contemporary urban soundscapes.

Without ignoring the fact that the sonic design of a specific site has a direct influence on an individual's health, the researchers had a keen eye on the socio-political role of sounds as related to a more general feeling of well-being. Sonically renegotiating and transforming (the experience of) a place is an affective process that is generated by the establishment of a myriad of affective lines between a range of heterogenic components that span between the human and the nonhuman, and the material and the discursive. Relevant issues that needed to be kept in mind: What new affective relations become possible with the suggested sonic interventions? Which other sounds should become audible or more dominant? And will this positively influence social behavior?

Friday, September 2022

The Hofplein Square or roundabout is situated right in the center of Rotterdam, in between Rotterdam Central Station and its most well-known street, the Coolsingel. Although the square has a rather big fountain as its main landmark, it is not an accessible place: getting near the fountain means you should not only cross a five-lane road but also make an effort to avoid getting hit by one of the hundred trams that traverse the square every hour. In short, the square is "owned" by motorized traffic. And wind. Therefore, although there are some places to sit and a few inter-esting sculptures by Rotterdam-born artist Willem de Kooning, the square currently functions only as a transition space.

In order to make this place more attractive and able to help mitigate urban heat stress and flooding, the city government plans to turn the Hofplein square into a space where shoppers and tourists can pleasantly spend some time and relax. This means, among several other things, turning the square into a low-traffic area, planting trees where concrete and pavement are now dominant, encouraging and facilitating the establishing of bars and coffee corners to replace offices around the square, and placing benches where now cars still rush past.

Of course, all these interventions directly affect the sonic environment. Instead of motorized traffic, sounds of humans (pedestrians and customers), trees, and birds should prevail in the future, although these will still face fierce sonic competition from trams, cars, signaling systems, and other extant electronic devices. The report that my research team and I wrote and submitted to the city government contained several recommendations to improve the place sonically as well as a list of potential pitfalls. One of the recommendations focused on the (aural) role of the fountain, which I will return to further on (Photo 9.2).

Photo 9.2 The Hofplein square with its fountain, Rotterdam. © Marcel Cobussen (2021)

9.4   The Role of Sound Art

Sound art increasingly engages with socio-political and ecological issues; artists explore a city's complex patterns, take into account everyday life practices, and negotiate between an aesthetization of the urban environment and its mundane functions. However, just as sound art in public urban spaces aims to transform the sonic—and thereby also the socio-political and ecological—environment, the already-existing urban ecology affects the ideas of the artist and the eventual artistic intervention (Groth and Samson 2013, p. 95).

There is a benefit to include at least one sound artist in the soundscape design team. In general, they are very good listeners, usually unprejudiced, and tolerant towards sounds that are frequently rejected by non-artists; not uncommonly, they ascribe a more affirmative significance to some existing, sometimes noisy, aspects of an urban environment. Additionally, they are able to contribute their own expertise and (aesthetic) ideas regarding the possible (re)design of a specific place, thereby renegotiating its already-established functions and organization.

The process of creating an artistic intervention in a public urban place can be considered as an extra or alternative mode of thinking about this place: on the one hand, such an intervention is created by exploring the heterogeneous and complex affective force relations which together constitute a place, and on the other it may act as a modification of those very relations through strategies of renegotiation, the production of new connections or the intensification of certain experiences, thereby influencing or transforming one's relation to this place. A sound artist thus discloses the complexity and potentialities of a place, which are typically hidden by the dominant and (thereby) familiar sounds of the everyday. Sound artists reveal what is present, yet missed, in most ordinary interactions with a site (Lacey 2016, p. 164). That is, while engaging with the power relations—including the already-existing sounds—that constitute a place, a sound artist will not stay with what a specific place already *is* but what it could *become*. They will experiment with and elaborate upon the inherent capacities of a place, and any transformation of those capacities might then lead to new affective experiences. In other words, an artistic intervention will consist of investigating, analyzing, and experimenting with the real capacities that are already actualized as well as those that have not yet been actualized—the not-yet-formed intensive forces of a place. Art can thus reactivate and expose the perhaps forgotten or suppressed possibilities of a site.

What becomes clear is that this type of sound art cannot be understood as an autonomous and isolated entity or work; it is a situated and social event, composed in dialogue with an environment that has its own specificities and characteristics. Elsewhere I have called this a *sonic ecology*: interactions between agents—in particular humans—and their environment based on sonic input, including music but also vibrations and resonances that are imperceptible to the human ear, such as infra- or ultrasounds (Cobussen 2016). Ecological sound art thus fosters a co-functioning or a concrete encounter between the artwork, the interacting human and nonhuman bodies, and the urban environment.[10] On the basis of a close analysis and investigation of the site's acoustics, visuals, architecture, materials, existing objects, function, meanings (historical, cultural, etc.), and perception, artworks should "grow out of the site, both conceptually and in their execution," thereby blending on various levels with the existing environment. Also playing a notable role in shaping the sound artwork: the passersby, natural elements (wind, temperature, humidity), and local everyday rhythms. So, besides transforming, sonic works may also intensify already-existing ecologies and reveal how heterogeneous components are intertwined and related (Groth and Samson 2013, pp. 102–103 and 110). Sound art may therefore encourage or refine our awareness of our surroundings, shifting sensibilities and deepening our relation to a specific place.

What does all this have to do with health? As discussed above, living bodies (humans, animals, plants, trees, etc.), non-human materialities, and the environment act upon one another and are acted upon. Put differently, all these bodies can be defined by their capacity to affect or be affected by other bodies. Following here is a concept of ethics—and not aesthetics!—as developed by the French philosophers Gilles Deleuze and Félix Guattari, partly based on their rereading of Spinoza's *Ethica*,

---

[10] Whereas site-specific artworks are primarily concerned with their formal attributes in relation to a concrete place, an ecological approach is more sensitive to the entanglements of various urban components and aims to enter into a dialogue with them, attempting to fluidly integrate the material and non-material environment on various levels while also involving human bodily behavior (Groth and Samson 2013, pp. 101–104).

the more a body is able to affect and be affected and the more a body can be attentive to new modes of affection, the better it is (Deleuze and Guattari 1994; Smith 1998). In other words, a healthy body is a body that is better able to realize itself in its full potential. And by diversifying sonic environments, by actualizing some of their affective potentials, sound art expands the possibilities for new modes of affection, new and creative encounters, so that new experiences can unfold.

Sound art in public urban spaces may be able to augment our experiences by challenging the constraining, dominant, monotonous, and repetitive mundane soundscape to which perhaps especially human bodies are exposed on a daily basis and which contributes to their alienation, their indifference or lack of commitment regarding their sonic environment (Lacey 2016, pp. 15–16). The contribution of sound art to create a healthier environment has to do with its capacity to decode, deterritorialize, and transform inherent spatial power structures in order to actualize the virtual capacities that are always already embedded within a site, waiting to be actualized. In these virtual capacities, aesthetics, ethics, politics, and healthcare converge in that they evoke "imaginative responses by diversifying the reductive affects of noisy soundscapes" (Lacey 2016, p. 160). And this is not necessarily achieved by reducing urban sounds, by finding quiet in the city, by beautifying the soundscape or by adding sounds of nature.[11] This can be realized, first of all, by expanding the affective potential of a given environment, thereby opening the possibility of reconfiguring one's relation to the sonic environment.

Friday, September 2022

Let's return to the Hofplein square and its fountain once more. The fountain currently has several nozzles which can all be turned on or off at the same time and with more or less the same strength throughout. A simple sonic analysis reveals that the fountain is either barely audible when there is high traffic or producing audible sounds that are barely distinguishable from traffic noise when there are only a few cars, trams, or motorcycles passing: a rather monotonous drone, similar to pink noise.

When the proposed and planned traffic reduction has moved from the realm of the virtual to the domain of the actual, the sound of the fountain will certainly become more prominent. From this arose the idea to do something with the fountain, to do something to make it sound more varied and therefore—hopefully and presumably— more interesting. At this time, the plan is still in the development stage, beginning with retrieving data about the technical infrastructure of the fountain and its nozzles. However, the idea—enthusiastically received by the project management—is to nominate a sound artist or a composer, commissioned by the City of Rotterdam, to create a sound composition for the fountain. The composition should consist of only (non-recorded) water sounds: by adapting and transforming the way the nozzles work,

---

[11] As sound scholar Jordan Lacey states, it is not so much the *stuff* of nature (actual sounds of birds or other animals) that needs to be recreated or implemented; what needs to be actualized is nature's *diverse affective potential* in the urban environment. The everyday environmental condition of the urban, the sonic homogeneity of urban noise, must be diversified in order to disclose new capacities and, in their wake, more healthy circumstances (Lacey 2016, p. 9 and 26).

they should be able to function independently from one another and vary the strength of their water jets. This will create a more heterogeneous sonic environment in which different sounds will become more prominent at different times: the water sounds, the sounds of the trees, the sounds of pedestrians and bar visitors, in interaction with the sounds of traffic, of cars, trams, bicycles, and planes. Ultimately, the work becomes part of the site's ecology; it brings with it its own specific agency and becomes connected to the ongoing activities of this square.

9.5   Listening

According to the Italian composer Luigi Nono, listening has the potential to put us in contact with "other thoughts, other noises, other sonorities, other ideas" that we were previously unfamiliar with and perhaps could not even have imagined before (Nono 2018). In this sense, listening can be regarded as an act of discovery. But, Nono warns, this listening to otherness, this other listening or listening differently, is not easy to achieve. Whereas Nono is ostensibly referring in this context to modern music and its often-proclaimed inaccessibility, my question here would be: Is this "discovering listening" somehow also applicable to our engagement with the everyday urban sonic environment? Are we able and willing to listen to other noises and other sonorities? Are we able and willing to suspend common prejudices against certain sounds, perhaps by not calling them "noise" in the first place?

Let me approach this from a different angle. I do agree with those who say that local, national, and international governments have the obligation to protect and/ or increase the well-being of their citizens, and one of those ways is by ensuring a healthy sonic environment. However, as we know already for a long time, the world is not (completely) malleable; not everything can be organized and regulated in such a way that everyone is always satisfied, happy, or feeling well. Finding a solution for unpleasant soundscapes can never be the responsibility of only urbanists, politicians, and project developers, as a successful top-down applied construct is not only an illusion but also politically and ethically inadvisable. Therefore, I would like to create an opening here for another approach as well, a plea to also take into account a bottom-up thinking, a thinking based on a more open and welcoming listening attitude as an alternative strategy for increasing one's well-being in a specific sonic environment.

In 1952, the composer John Cage shocked the world of classical music with his composition *4′33″* which consisted of a performer *not* playing their instrument and (thus) remaining relatively silent for over four and a half minutes. Cage's intention was not only to make people aware that silence doesn't actually exist—we are always surrounded by sounds; equally important was Cage's artistic statement that each sound or sounding constellation can be considered as music. For Cage, listening to a rather random soundscape was often more interesting and exciting than attending a Beethoven symphony, as it was refreshingly unpredictable, complex, and stimulating one's imagination. As he explained in his book *Silence*: "Wherever we are, what we hear is mostly noise. When we ignore it, it disturbs us. When we listen to it, we find it fascinating. The sound of a truck at fifty miles per hour static between the stations. Rain" (Cage 1973, p. 3). What Cage was asking from his audience was whether they could listen to those everyday sounds *as if* they were music, thereby emancipating both the classical music world and sounds previously not considered as music.

Whereas Cage brought "non-musical" sounds into the concert hall, sound artist Max Neuhaus didn't feel the need to enter such a hall anymore. Between 1966 and 1968, he organized several soundwalks, taking the audience outside to experience the acoustic everyday world itself. He asked them to gather at a particular spot on the street, put a stamp with the word LISTEN on their hands, and led the group through their everyday environment. The idea was for people to concentrate in silence on the listening experience, thus recalibrating their auditory perspective. The Japanese artist Akio Suzuki creates some of his performances in a more or less similar way. *oto-date* is a series of works in which Suzuki uses only the existing sounds of a city. On specific spots he paints ears that resemble feet on the ground; by positioning yourself on those ears, you listen to the environment from a place which was deemed particularly interesting by Suzuki. With *oto-date*, Suzuki attempts to disclose usually-ignored sonic peculiarities in our everyday sonic environment, an environment full of sonic mysteries, if only we care to listen (Lacey 2016, p. 167) (Photo 9.3).

Whether through in-person interactions with sound artists or other composers of soundwalks or not, works like those of Cage, Neuhaus, and Suzuki invite people to engage with their sonic environments in a critical way. Or as sound artist Chris Watson put it: "The whole aim was to make [...] people think about the sounds of their city" (Watson in Gandy and Nilsen 2014, p. 167). Only by creatively and actively listening, can people slowly start to sense what they like and what they don't like. On the one hand, this may lead to more acceptance of the soundscape as it is (or at least parts of it) and a sensitivity towards protecting certain sounds that give the environment its specific character. On the other hand, this open and attentive listening attitude can be a first step towards reflecting on how displeasing sounds or acoustics should be changed.[12] Both serve the same goal: to create a sonic environment that contributes to the general well-being of all human and nonhuman beings living or spending time there.

---

[12] It is important to stress that this attentive listening should not be equated with being able to cognitively *understand* sound or a sonic environment. Attentive listening certainly also implies becoming aware of one's affective relations (or not) to a site, relations that are as much embodied as they are established by the mind or factual knowledge. Attentive listening means to become aware of the urban soundscape's ability to shape the physical and emotional expressions of the collective social body.

Photo 9.3   Akio Suzuki's *oto-date*. With permission

9.6   Conclusion

Sounds can affect people's health and well-being in a negative but certainly also in a positive way. Especially when dealing with sounds in public urban spaces, the emphasis has often been on the negative impact: sound is frequently equated with noise. In this chapter, I have tried to focus more on the potentialities of sound to positively contribute to the overall health of all those interacting within the sonic ecology of an urban site and thereby to a more general and social well-being.

With the section on listening, I have tried to make clear how any intervention in a sonic environment should begin with listening experiences rather than applying so-called objective measurements or uncritically complying with laws and rules which usually only consider loudness. Furthermore, the responsibility for one's well-being cannot one-sidedly rest in the hands of acousticians or local, national, or international authorities; lending an ear to residents, users, and passersby is crucial to improving a sonic environment. And although the idea of improving is often interpreted to mean adding more natural sounds to urban sites, I have argued that (re)designing for social health and well-being also means encouraging people to explore, engage, play, and wonder. In other words, sonic interventions should not only, and even not primarily, aim at making a site more beautiful (whatever that would mean); their main objective should be to enlarge affective capacities, to enhance the possibilities for human as well as non-human agents to affect and be affected, for example by moving beyond the familiar and thereby transforming the experience of a particular site. My claim is that sound artists are perfectly equipped to help reshape and expand those affective capacities, also within the context of the everyday.

Sunday, March 2023

Dear reader, you have come to the end of this essay, which has laid a claim mainly on your eyes and mind. But your ears were not covered; consciously or unconsciously, they have registered many different sounds. And, consciously or unconsciously, they have affected your mood, your concentration, your relation to this text. Consciously or unconsciously, positively or negatively, by affecting your mind as well as your body, your sonic environment (co-)determines your well-being. Simply becoming aware of the working of these sounds, of listening to them and exploring your relationship to them, is a first step. A second step is to investigate them further: which of these sounds are worth listening to? Which sounds would contribute to a more general well-being, both on an individual and a social level? How do sounds combine to create a space in which I feel safe and pleasant? The third step would be to concretely intervene in your sonic environment so as to increase your capacities to affect and be affected.

References

Cage J (1973) Silence: lectures and writings by John Cage. Wesleyan University Press, Hanover

Cobussen M (2016) Towards a "New" sonic ecology. Inaugural lecture. Leiden University, Leiden

Cobussen M (2022) The sonic turn: towards a sounding sonic materialism. New Sound 60/II:11–24

Deleuze G, Guattari F (1994) What is Philosophy? (trans: Tomlinson H, Burchell G). Columbia University Press, New York

Epstein MJ (2020) Sound and noise: a listener's guide to everyday life. McGill-Queen's University Press, Montreal

Gandy M, Nilsen BJ (eds) (2014) The acoustic city. Jovis, Berlin

Goodman S (2010) Sonic warfare. Sound, affect, and the ecology of fear. MIT Press, Cambridge

Groth SK, Samson K (2013) Urban sound ecologies. an analytical approach to sound art as assemblage. Sound Effects 3(3):95–112

Johnson B (2023) Earshot. Perspectives on sound. Routledge, New York

Lacey J (2016) Sonic rupture. A practice-led approach to urban soundscape design. Bloomsbury, New York

Nono L (2018 [1983]) Error as necessity. In: De Benedictis A, Rizzardi V (eds) Nostalgia for the future, Luigi Nono's selected writings and interviews. University of California Press, Berkeley, pp 367–369

Sennett R (2018) Building and dwelling: ethics for the city. Farrar, Strauss and Giroux, New York

Smith DW (1998) The place of ethics in Deleuze's philosophy. In: Kaufman E, Heller KJ (eds) Deleuze and Guattari: new mappings in politics, philosophy, and culture. University of Minnesota Press, Minneapolis, pp 251–269

Thibaud J-P (2003) The sonic composition of the city. In: Bull M, Back L (eds) The auditory culture reader. Berg, Oxford, pp 329–341

Zimmer C (2019) The sounds that haunted U.S. Diplomats in Cuba? Lovelorn crickets, scientists say. New York Times, January 4

Chapter 10
Future Developments in Noise from Transport

Antonio J Torija Martinez

Abstract The world is currently undergoing a significant transition towards cleaner and more sustainable energy sources. The transportation sector is gradually moving away from fossil fuels and electric vehicles, both on the ground and in the air (e.g., drones), are more and more common. The introduction of these electric vehicles will bring new sources of transportation noise, which might lead to the largest shift in soundscapes in living memory. This soundscape shift could be detrimental to the public health and well-being if appropriate actions are not taken. This chapter presents the state-of-the-art of the fast-developing field of transportation noise, and discusses current practice gaps and recommendations.

We need to start imagining (and asking ourselves) what the future is going to sound like above and under water,

What do we want our future to sound like and how do we get there?.

Spence in https://planetforward.org/story/marine-ecologists-sound-pollution/

10.1 Introduction

The world is currently undergoing a significant transition towards cleaner and more sustainable energy sources. During this energy transition, there is a gradual move away from fossil fuels and an increased reliance on renewable energy technologies such as wind, solar, and hydroelectric power. This energy transition is expected to bring substantial environmental and socioeconomic benefits; but it is important to also account for the impact of the noise generated by these renewable energy installations on human' and wildlife's health and well-being.

Wind turbine noise has been a focus of environmental noise research for several years (Hansen and Hansen 2020). The impact underwater noise produced by offshore

A. J. Torija Martinez (✉)
Acoustics Research Centre, University of Salford, Greater Manchester, UK
e-mail: A.J.TorijaMartinez@salford.ac.uk

© The Author(s) 2025
I. van Kamp and F. Woudenberg (eds.), *A Sound Approach to Noise and Health*,
Springer-AAS Acoustics Series, https://doi.org/10.1007/978-981-97-6121-0_10

wind farms has on wildlife has been investigated. For instance, Madsen et al. (2006) reported that high sound levels during construction activities are likely to disrupt the behaviour of marine mammals at ranges of many kilometers. However, Mooney et al. (2020) suggest that further research is needed to have a comprehensive understanding of the effects of offshore wind farm noise on wildlife. The noise generated by wind turbines is frequently a cause of complaints from communities living near wind farms due to noise annoyance and sleep disturbance (Nguyen et al. 2021). The noise annoyance due to wind turbine noise has been usually associated with several acoustic features, such as the presence of infrasound, a low-frequency dominated spectrum (Zajamšek et al. 2016), tonality (Liu et al. 2012), and amplitude modulation (Nguyen et al. 2021). Noise annoyance due to wind turbine noise is correlated to sound levels; but is also associated with several non-acoustic factors, e.g., both objective and subjective factors of wind turbine visibility (Pedersen and Waye 2007). Due to these acoustic and non-acoustic factors, some studies (Pedersen and Waye 2004) have found wind turbine noise to lead to a higher percentage of highly annoyed people than expected from the existing dose–response relationships for transportation noise (Miedema and Oudshoorn 2001).

Decarbonising heating and cooling is one of the main goals of the European Environment Agency (EEA 2023). Heat Pumps have been suggested as a key technology for the decarbonisation of heating in households. In the UK, the Government's Net Zero agenda is planning a wider deployment of Heat pumps, at a rate of 600,000 a year from 2028. However, these technologies do not come without drawbacks and challenges. Noise has been regularly suggested as one of the main barriers to the wider adoption of heat pumps. Some of the acoustic features of heat pump noise have been comprehensively studied, such as vibration-induced noise, low-frequency noise, and tonal noise (Waye and Rylander 2001; Yonemura et al. 2021). There are also some important challenges still to be further investigated, such as how communities will respond to a sound environment with multiple heat pumps operating under different regimes; and what the contribution is of transient behaviours (e.g., de-frosting) on noise annoyance. This further research is a key priority of the working group Annex 63 of the Heat Pump Technologies (HPT) Technology Collaboration Program (TCP) of the International Energy Agency (IEA).

The transportation sector is also in a process of transition towards more electric and autonomous technologies. Transportation noise is usually reported to be the most important source of environmental noise (Clark and Stansfeld 2007). Therefore, the remaining of this chapter focuses on expected developments in transportation, and their implications on environmental noise and its effects.

This chapter presents the state-of-the-art of the fast-developing field of transportation noise, and discusses current practice gaps and recommendations.

10.1.1 Transportation Noise: Towards Electric Mobility

The soundscapes in which we live and work affect us in several ways, every moment of every day; and these soundscapes are expected to change dramatically in the coming years, whether we like it or not, as part of a major shift towards electric-driven mobility. Imagine a city in 2030, electric vehicles have taken over and the sky is inundated with drones and other novel aircraft; on the ground, electric vehicles and two-wheeled transport are dominating (see Fig. 10.1).

Electric mobility (or e-mobility) will lead to vehicles with entirely new sound sources. On the ground, the move away from internal combustion engines and towards e-drives would, in principle, lead to quieter vehicles as engine noise will be significantly reduced. However, a noticeable reduction in the overall noise reduction of Electric Vehicles (EV), compared to combustion engine cars, happens only at low speeds (i.e., lower than 30 km/h) where engine noise is dominating (Iversen et al. 2013). The overall noise reduction at higher speeds is less certain, and even a small increase in rolling noise caused by tire-road contact might happen due to an increase in EV weight consequence of carrying heavy batteries. Even with a quieter EV, this could be actually more annoying than a louder combustion engine vehicle, partly because the EV is different in noise spectrum and character (e.g., more high-frequency noise), but also because the quieter e-drive can reveal other vehicle sounds which were previously masked (e.g., tonal noise).

Fig. 10.1 Illustration of an urban scene with electric scooters and drones flying over. *Image generated by Midjourney [Large data model], (2024) from A man riding a scooter, by Pony (@getapony), 2022. (*https://unsplash.com/photos/a-man-riding-a-scooter-OHxsu4HTz5c*). Unsplash licence*

On the other hand, quieter EVs at low speeds could go undetected, and probably form a risk for pedestrians nearby. Therefore, these EVs must generate sound artificially to alert other road users, with potentially non-harmonious consequences for local communities. If not properly designed, this mixture of artificial alert sounds from different vehicles could be a factor of significant community noise annoyance.

In the air, drones (or other novel aircraft such as electric Vertical Take-Off and Landing—eVTOL—vehicles) will bring unconventional noise signatures. In these vehicles, the sound will be eminently tonal and high-pitched (Torija and Clark 2021). Tonal noise has been found to be strongly associated with noise annoyance, while high-frequency content has been found as one of the most important contributors to aircraft noise annoyance (Torija et al. 2019). There is enough evidence to suggest that the sounds of these novel air vehicles do not resemble the sounds of conventional aircraft (Christian and Cabell 2017). Neither will be the operating characteristics. Drones and eVTOLs will operate closer to communities (than conventional aircraft), and over urban (and possibly rural) communities not usually exposed to aircraft noise. All these new sources will certainly lead to the largest shift in soundscapes in living memory.

This soundscape shift could be detrimental to the public health and well-being if appropriate actions are not taken. However, there is also a scenario where drones move rapidly and quietly through the air; and electric surface transportation provides a pleasant background hubbub. To do this, manufacturers and decision-makers need the tools for carefully designing the sound of e-mobility vehicles so that they produce an optimal sound, taking citizens' requirements into consideration.

To realise this scenario, new perceptually driven engineering methods are needed. The concept of perception-influenced (or perceptually-driven as referred to in this chapter) engineering was first introduced by Davies and colleagues at Purdue University, to integrate the ways people perceive, or are affected by, machinery outputs into the design of engineered systems (Davies 2007). These perceptually-driven methods allow putting the public at the centre of engineering decisions to ensure responsible innovation. With these perceptually driven methods, manufacturers could listen to the effects of early design changes in their prototypes, and optimise the product sound for the user and their environment. This would allow manufacturers to fully realise the benefits of industrial strategies, such as Industry 5.0 in the European Union (Cotta and Breque 2021), pushing for a translation to a sustainable and human-centric industry.

The challenges are several and complex, including:

- A better understanding of the noise emission characteristics of e-mobility vehicles (as compared to their equivalent ground and aerial vehicles);
- New or updated sound emission and propagation models able to account for the unconventional noise signatures and operating conditions of e-mobility vehicles;
- Psychoacoustic knowledge to understand the human response to the sound generated by e-mobility;
- New or updated policy and guidance to inform vehicle and operation development to limit the impact of these new sound sources on communities.

But, at the same time, the introduction of e-mobility provides an excellent opportunity to change the way we have traditionally addressed the problems of environmental noise, and therefore allows us the opportunity for a fresh start to shape future soundscapes the way citizens want.

10.2 Drones and Other Novel Aircraft as New Sources of Environmental Noise

Several recent studies have found drones reported to be more annoying than other transportation vehicles, at the same sound level. A pioneering study by Christian and Cabell (Christian and Cabell 2017) compared the annoyance of a series of drone flyovers with road vehicles passing-by. They found the drones evaluated (with the number of rotors varying from 4 to 8, and weight from 1.6 to 8 kg) to be equally annoying as road vehicles with a 5.6 dB higher sound level; in other words, road vehicles had to be 5.6 dB louder to be perceived as equally annoying as drones. The authors hypothesised that this offset in annoyance is due to the specific sound characteristics of drones (i.e., tonal and high-frequency noise), and also due to the different flight operations (e.g., flying closer to people). Similar findings have been found by other researchers. Torija and Li (2020) found a small quadcopter 33% less preferred than a conventional civil aircraft taking-off (at the same sound level, 65 dBA); Gwak et al. (2020) found hovering drones equally annoying as a jet aircraft taking-off with a 4–10 dB higher sound level, depending on the size of the drone. It should be noted that drone and propeller technology is advancing rapidly, so these offset values might be soon obsolete and further research would be required.

There are several reasons why drones are more annoying than other transport vehicles at the same sound level, related to the 'sound signature' of drones. To start with, the concentration of acoustic energy in the high-frequency region (see Fig. 10.2) is one of the main differences between the noise signature of drones and other conventional civil aircraft (Gwak et al. 2020). The sound produced by a drone is very particular. In the case of multirotor drones, the propellers usually rotate at slightly different velocities which causes the presence of a multitude of discrete tones at specific frequencies. This makes the sound of a drone highly tonal; but also 'rough' as the multitude of discrete tones can interact with each other leading to fast modulation phenomena (equivalent to the sound of a 'sporty' car). The interaction between rotors, and between rotors and fuselage, produces unsteady pressure fluctuations causing high-frequency noise (Hubbard 1991). The operation of electric motors also produces the generation of high-frequency noise (Cabell et al. 2016).

Drone noise is also highly influenced by ambient weather conditions (Alexander et al. 2019). The flight control system of a drone varies individual rotor speeds to maintain vehicle stability, and creates an unsteady noise signature with rapid temporal fluctuations of the tonal components. Small variations in the frequency of the different rotors lead to large variations at higher frequencies (see Fig. 10.3). Together with

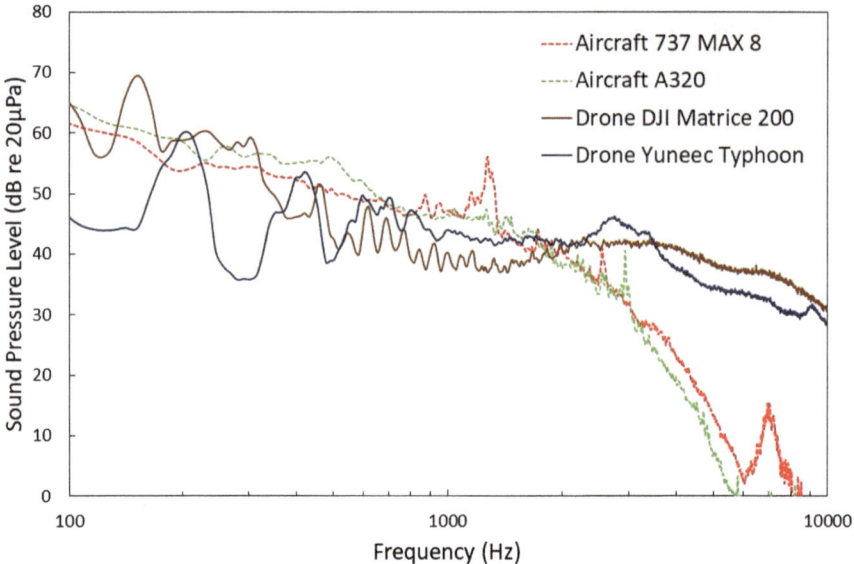

Fig. 10.2 Frequency spectra of two civil aircraft (Airbus A320 and Boeing 737-8MAX) and two small multi-copters (DJI M200 and Yuneec Typhoon), with an overall sound pressure level normalised to 65 dB(A) for comparison. Modified from (Torija and Clark 2021), licensed under CC-BY 4.0

the high-frequency sound of the motor, this creates a very noticeable high-frequency sound.

Another reason why drones are more annoying is that drones operate in a significantly different manner to conventional civil aircraft and on most occasions over

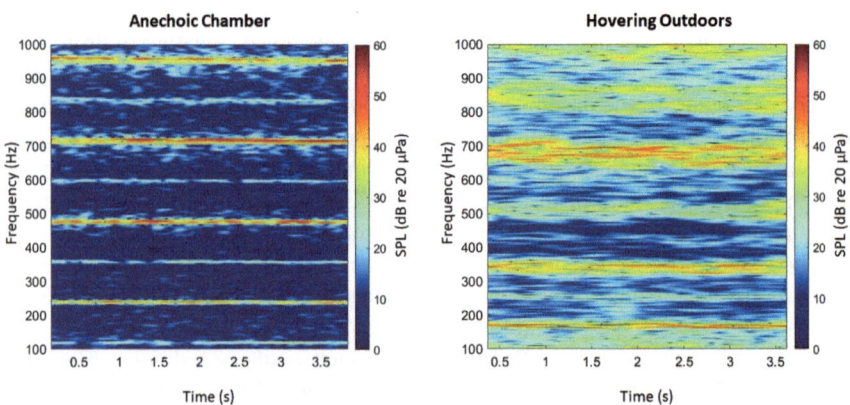

Fig. 10.3 Spectrogram of a DJI Phantom 2 quadcopter measured in an anechoic chamber (left) and measured outdoors while hovering (right). Modified from (Torija et al. 2019), licensed under CC-BY 4.0

communities not currently exposed to aircraft noise. In conventional civil aviation operations, flight profiles are designed to quickly move aircraft far away from exposed communities. Thus, in communities living around airports, aircraft height about ground would be about 6,500–7,500 ft (around 2–2.3 km). Drones will operate much closer to exposed communities, i.e., not higher than 400 ft (i.e., 120 m) above the ground. In a typical operation for a parcel delivery, a drone would approximate the property of destination, would descend, stay hovering for several seconds, then ascend and fly away again. This implies that the drone operation close to citizens can lead to noise annoyance. Christian and Cabell (2017), suggest that a 'loitering' penalty would account for some, if not all, the differences in noise annoyance between drones and road vehicles.

In summary, noise annoyance from drone operations has been found to be primarily influenced by how loud the sound is perceived, the presence of high frequency (or high pitch) noise, and the presence of amplitude-modulated sound due to the interaction between rotors (Gwak et al. 2020; Torija and Nicholls 2022); and the presence of tonal noise (Torija and Li 2020).

10.2.1 Urban Air Mobility

A new aviation sector is also expected to expand in the next few years: Urban Air Mobility. The main motivation here is to contribute to a multimodal mobility system, enabling the exploitation of urban skies for people's transportation. Building upon the ongoing development of electric powertrains and battery technology, a new generation of aircraft is under research and development. These novel aircraft include several configurations, although the main designs pivot around eVTOL vehicles. Most of these eVTOLs are based on multi-rotor configurations, which produce a noise significantly different from the conventional rotorcraft and propeller-driven aircraft. As for the drones, these novel aerial vehicles bring significant acoustic challenges due to their unconventional noise signatures, with more tonal, high-frequency broadband and time-varying noise; and also unconventional maneuvers such as the transition from hover to forward flight. If not appropriately considered and managed, these noise emissions and operating characteristics will likely lead to important problems of environmental noise.

10.3 Change in Soundscape with e-Mobility

The transition to electric mobility could have detrimental effects on the soundscape of cities, such as a shift towards high frequencies which are usually perceived as more annoying and unpleasant. Urban soundscapes are currently dominated by road traffic noise, which has been traditionally generated by fossil fuel or internal combustion engine vehicles. There is a significant difference between internal combustion engines and electric powertrains and so their sound generation mechanisms are different. The sound emission of electric powertrains can be up to 20 dB lower (in A-weighted sound level) in full acceleration mode than conventional internal combustion powertrains. However, their sound signature is dominated by high frequencies and tonal components in the frequency range from 1 to 10 kHz (Muender and Carbon 2022). The human auditory system is particularly sensitive to these high frequencies. By being very quiet, and with the absence of the typical broadband noise spectrum of internal combustion engine vehicles, other disturbing noises are unmasked. These include switching noise caused by power electronics, with frequencies ranging from 250 Hz to 20 kHz, which has been found to be experienced as quite unpleasant. For the specific case of vehicle interior sound quality and comfort, electric powertrains are potentially more annoying than internal combustion powertrains due to their acoustic profile with higher frequencies and tonal components (Muender and Carbon 2022; Lennström and Nykänen 2015; Swart et al. 2016; Lennström et al. 2013).

A literature survey about noise from electric vehicles (Marbjerg 2013) described the frequency content of noise from electric vehicles under different speeds, and for different electric and hybrid electric vehicles. The common finding was that the frequency spectra of electric vehicles have much less content in low-frequency noise, and much more content in high frequency noise. For instance, Fig. 10.4 (modified from (Wachter 2009)) shows sound levels of electric vehicles at frequencies between 1 and 2 kHz higher than the sound levels of internal combustion engine vehicles at a speed of 50 km/h.

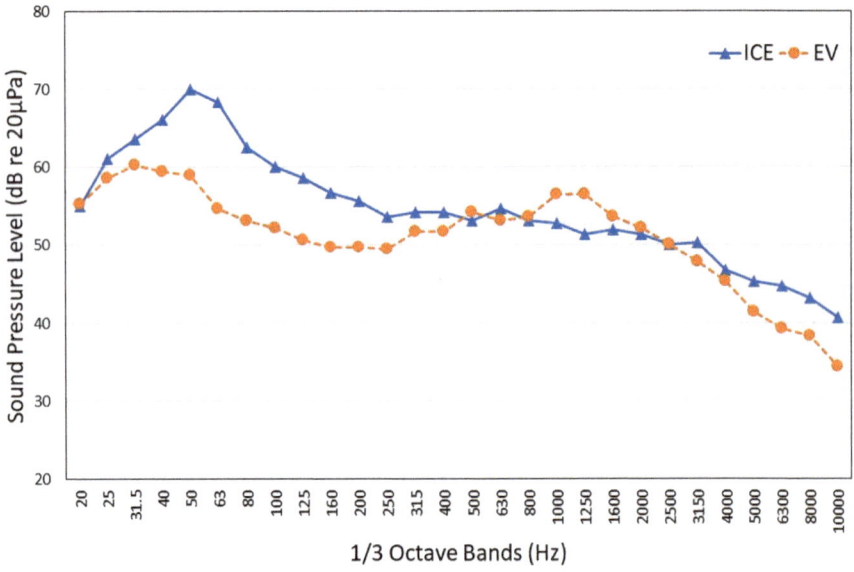

Fig. 10.4 Frequency spectra of an internal combustion engine (ICE) vehicle and Electric vehicle (EV) at a constant speed of 50 km/h. Modified from (Wachter 2009)

The literature on the perception of noise from electric vehicles is scarce and focuses mainly on vehicle interior sound quality. A study conducted by Govindswamy and Eisele (Govindswamy and Eisele 2011) investigated what parts of the frequency spectrum are more important for sound perception in electric vehicles from the driver's perspective. Using an electrified version of the Fiat 500, and varying the sound level at different frequency regions,[1] the authors found that: (i) reducing the low-frequency content had no effect on noise perception ratings; (ii) reducing the mid-frequency content improved the reported pleasantness and dynamic impression rating; and (iii) reducing the high-frequency content lead to the biggest improvements in reported pleasantness and preference. Similar findings were reported by Lennström et al. (Lennström et al. 2011). In their sound quality evaluation of EVs in vehicle's interior, the authors found that increasing the sound levels of tonal components at high frequencies led to high values of reported sharpness (i.e., sensation based on the amount of high pitch noise), annoyance, toughness/aggressiveness, and powerfulness; while a reduction of sound levels of tonal components at high frequencies yielded high rankings in the overall satisfaction of the sound produced by the EV. The understanding of changes in environmental noise perception with the introduction of EVs is rather limited and must be further investigated.

Regarding changes in noise pollution in the environment, the introduction of e-mobility vehicles can lead to an overall reduction of sound levels, as EVs are significantly quieter than internal combustion engine vehicles (Marbjerg 2013). A report by the National Institute for Public Health and the Environment in the Netherlands (Verheijen and Jabben 2010) estimated a fully electrified fleet of road vehicles to lead to an overall reduction of sound level between 3 and 4 dB. The main assumption was that 90% of passenger cars and light freight cars, and 80% of heavy trucks were electric. The report also suggested that the largest reduction of 4 dB will be on secondary urban roads and intersections (with lower average speeds); and also that for speeds above 50 km/h, EV and hybrid vehicles were not quieter than conventional internal combustion engine vehicles. This is because rolling noise (or tire-road noise) is the dominant noise source at high speeds, in contrast with engine noise which is dominant at low speeds. Campello-Vicente et al. (Campello-Vicente et al. 2017) investigated the effect of the replacement of internal combustion engine vehicles with EVs on the overall sound levels presented in strategic noise maps. The authors also considered the effect of the Acoustic Vehicle Alerting System (AVAS) on the overall sound emission. Assuming all passenger cars to be electric and no heavy vehicles in traffic, and an average speed of 30 km/h in a free field lane, an overall sound level reduction of 2 dB was found. This overall reduction of sound level dropped to 1 dB, if the use of the AVAS in electrics passenger cars was assumed.

With a decrease in noise in urban environments due to EVs replacing internal combustion engine vehicles, other noise events can become more noticeable, and therefore, lead to an increase in community noise annoyance. This is the case with novel aircraft concepts operating in the skies of our cities, such as drones. A study

---

[1] The authors presented the original recording of the electrified version of the Fiat 500 vehicle, and also the original recording with the different part of the frequency spectrum attenuated.

based on laboratory simulations found that at locations with dominant road traffic noise, the presence of a drone led to an increase in the reported annoyance 1.3 times the annoyance without the drone; while at locations with low road traffic noise, the annoyance due to the presence of the drone increased 6.4 times (Torija et al. 2020). In these locations with low traffic noise, the noise annoyance was always about 7 (on a scale from 0 to 10), regardless of the overall A-weighted Energy Equivalent Sound Pressure Level (L_{Aeq}) in the location. This suggests that the L_{Aeq} is not an appropriate metric for assessing the annoyance due to drone operations. Since A-weighted time-integrated sound level metrics are widely used in environmental noise mapping, assessment and planning, this is an important finding. LAeq based metrics may not be the most appropriate ones to assess new urban soundscapes with more clearly noticeable sequences of noise events, both on the ground and in the air. Other metrics might be worth consideration, such as the Intermittence Ratio (Wunderli et al. 2016) accounting for the acoustic energy contribution of individual noise events above a threshold; or other acoustic or psychoacoustic metrics accounting for unconventional frequency and temporal characteristics.

10.4   Challenges in AVAS for e-Mobility

As discussed above, EVs might be almost silent at low speeds (i.e., below 30 km/h) due to the significant reduction of mechanical sounds produced by the vehicle powertrain. Although this can lead to a reduction of environmental noise, and therefore, minimise adverse health outcomes due to noise exposure (Campello-Vicente et al. 2017), it could pose a growing threat to pedestrians (and other users of the public space) in the form of collisions. Several associations of the blind and visually impaired, including the Royal National Institute of Blind People in the UK, have advocated for the addition of artificial acoustic signals to increase the detectability of EVs. The hazard of 'near-silent' EVs has been widely featured in mass media (Fiebig 2020).

The US National Highway Traffic Safety Administration (Hanna 2009) found increased incidence rates of pedestrian and bicyclist accidents where EVs were involved, compared to internal combustion engine vehicles. The study also found an increased risk of collisions with EVs (compared to internal combustion engine vehicles) for visually impaired people. Karaaslan et al. (Karaaslan et al. 2018) found that the risk of road traffic near-misses and accidents involving pedestrians was around 25% more likely when comparing EVs with no AVAS to internal combustion engine vehicles. Several near misses have been recorded in Norway involving pedestrians with impaired vision when crossing roads. A third of the members of the Blind Union of Norway now say they are more afraid to move around in traffic (Berge 2018). Other studies in Norway also found EVs are more likely to collide with pedestrians and cyclists than internal combustion engine vehicles, possibly because of the low noise levels (Liu et al. 2022).

Electric scooters (or e-scooters) are now a common form of transportation in cities, with an estimated number of e-scooters across Europe of 520,000 in 2022. A study by the UK Department for Transport on road traffic collisions involving e-scooters found an increase from 484 in 2020 to 1,356 casualties in collisions in 2021 (DfT 2021). In a survey on perceptions of current and future e-scooters used in the UK (KANTAR 2021) 53% of the respondents suggested safety issues as one disadvantage of these vehicles.

These issues with the safety of pedestrians, and other users of the public space, including the blind or visually impaired, have led to the development of regulation for the design and use of AVAS in EVs. Currently, there is a range of regulations specifying the requirements of AVAS for EVs (Fiebig 2020). For instance, the United Nations Economic Commission for Europe (UNECE) Regulation 138 specifies the minimum required sound levels in one-third octave bands between 160 Hz and 5 kHz, and states that complying with alerting sounds requires minimum levels in at least two of the specified bands and with one of them below or within the 1600 Hz one-third octave band. There is a good degree of agreement between different regulations, although there are some differences. For instance, European regulation (No 138 of UNECE, U.E.R. 2017) requires the AVAS to operate up to 20 km/h, and include a pitch shifting with speed (not mandatory in US regulation); while the US regulation (NHTSA 2016) requires the AVAS to operate up to 30 km/h, and produce an alerting sound while the vehicle is stationary (not mandatory in European regulation).

To date, there is no regulation requiring AVAS for e-scooters, and therefore, there is no guidance on specifications of alerting sounds for these small vehicles. However, the UK Government has recently suggested e-scooter audibility as a key point to be included in future policy development (DfT 2022). To this end, Torija et al. (2023) and Walton et al. (2022) have conducted research into the detectability of e-scooters in a range of environmental noise conditions to aid the development of AVAS for micromobility transport.

The design of alert sounds for AVAS needs an appropriate consideration of the balance between detectability and annoyance. In other words, manufacturers want their vehicles to sound distinctive and identifiable, but do not want their vehicles to be associated with annoying sounds; at the same time, regulators want EVs to be detected to avoid risks of collision with pedestrians and other users of the public space, but do not want these vehicles to contribute to noise pollution in cities.

The addition of pure tones, and amplitude modulation and impulsive characteristics seem to be beneficial for increased detectability and localizability. A problem arises when a fleet of EVs of different types, and producing different alerting sounds operate at the same time and location. Each type of vehicle should have an appropriate alerting sound, that in addition to comply with regulation if existing, allows pedestrians to clearly associate the sound with the vehicle. For instance, the characteristics of the sound of an e-scooter (e.g., pitch) should be recognisable as a sound produced by a small vehicle operating at low speed (below 20 km/h), and therefore cannot be the same as the alerting sound for an electric bus.

Superposed alerting sounds, with different pitch, pitch-shift factor, and noise character can lead to dissonant and inharmonious urban soundscapes (Laib and

Schmidt 2019). Soundscapes composed of dissonant sound patterns caused by several 'untuned' superposed alerting sound signals could increase noise annoyance. Therefore, assuming a transition towards a fleet of electric vehicles operating in urban settings, from e-scooters to electric trucks, avoiding unintended effects such as an overall increase of noise annoyance due to different AVAS would require a close alignment between regulations for different types of vehicles, but also comprehensive studies investigating the acceptance of soundscapes with a range of AVAS in operation.

10.5 Research and Policy Gaps

10.5.1 Drones and Other Electric Novel Aircraft

In 2020, the NASA Urban Air Mobility (UAM) Noise Working Group published the white paper *'Urban Air Mobility Noise: Current Practice, Gaps, and Recommendations'* (Rizzi et al. 2020). Although the focus of the white paper was UAM vehicles, i.e., aircraft for public transportation in urban settings, part of the gaps and recommendations are also of application for small to mid-size drones (i.e., below 600 kg of total weight including payload). This white paper overviews the current practice, identifies gaps, and makes recommendations in four areas of interest: (1) tools for acoustic prediction; (2) ground and flight testing; (3) human response and metrics; and (4) regulation and policy. Areas (3) and (4) are of more interest for this book chapter.

There is currently some regulation and guidance on drone noise measurements, for instance, the Commission Delegated Regulation (EU) 2019/945 of 12 March 2019, amended by the Commission Delegated Regulation (EU) 2020/1058 of 27 April 2020. This regulation requires the calculation of Sound Power Level (L_W) for drones in the 'Open Category' to be measured during hover above one reflecting (acoustically hard) plane, according to EN ISO 3744:2010. Regulation 2019/945 also includes maximum Sound Power Level requirements, as a function of the weight of the drone (always below 4 kg). For outdoor conditions, other guidance currently in place includes: 'Guidelines on noise measurement of Unmanned Aircraft Systems lighter than 600 kg operating in the specific category' developed by the European Union Aviation Safety Agency (EASA), the 'NASA UAM ground and test measurement protocol', and the ISO 5305:2024—Noise measurements for UAS (unmanned aircraft systems). These guidelines specify detailed methods for an accurate characterisation of the noise produced by drones under actual operating conditions outdoors. However, what these guidelines do not include are noise limits for drone operations, as they are set for other aircraft and rotorcraft.

The lack of noise limits for drone operations is probably due to the scarce evidence on human response to drone noise. Although the evidence of drone noise effects on humans is very limited, some conclusions can be drawn from the literature (Schäffer et al. 2021). For instance, drone noise is reported to be more annoying than road traffic and aircraft noise (at the same sound level) due to particular acoustic characteristics such as the dominant presence of tonal and high-frequency noise. However, other factors such as the influence of factual and situational context, existing soundscape and audio-visual interactions on noise effects of drones have not been explored to date. The need for further research to better understand the effects of drone noise on exposed communities includes the development of noise metrics to assess the community noise impact of drones; the definition of acceptable levels for drone noise; the development of noise abatement procedures for drone operations; and the innovation in approaches to predict the long-term effects of drone noise exposure (Torija and Clark 2021). The latter is of particular importance, as it will allow to define exposure–response relationships for drone noise, as a key to carrying out an appropriate management of the noise produced by drone operations.

10.5.2 Electric Ground Mobility

There are two important issues associated with the replacement of internal combustion engine vehicles with EVs: (1) the shift in the frequency spectra of EVs toward higher frequencies (compared to internal combustion engine vehicles), leading to a potential increase in noise annoyance and (2) the need to add artificial alerting sounds to EVs to enhance noticeability at lower speeds, potentially increasing noise annoyance due to the use of acoustics features such as pure tones (at relatively high frequency) and amplitude modulation.

Some research has been done to be able to tackle these issues (Pallas et al. 2016). Further research is required, but this is not a simple task due to the change in the contribution of dominant sources (at different speeds) compared to internal combustion engine vehicles, the uncertainty of differences in rolling noise in EVs compared to internal combustion engine vehicles (Marbjerg 2013), and the quantification of the contribution of artificially added alerting sounds to the overall noise emission of an EV. In addition to this, it is unknown how communities will respond to a soundscape composed of a multitude of several alerting sounds with different characteristics. Comprehensive research is needed to better understand the potential change in noise perception of road traffic when conventional low-frequency propulsion noise is replaced by alerting sounds using tonal, amplitude modulation, and other acoustic features to increase the noticeability of EVs.

Another issue to address is the lack of regulation for artificial alerting sounds for micromobility (i.e., electric scooters). As for electric cars, specific requirements for the acoustic features of alerting sounds in e-scooters are deemed necessary for vehicle manufacturers to ensure that their vehicles do not create a risk for pedestrians and other users of the public space. The expectation is to provide minimum requirements of sound emission, frequency content, temporal characteristics, and directivity to ensure an appropriate balance between maximum noticeability and minimum noise annoyance. The use of psychoacoustic methods as suggested by

Fiebig (2020), or implemented by Walton et al. (2022) should allow the careful design of alerting sounds including key acoustic features to increase vehicle detectability for pedestrians without necessarily leading to an increase in noise annoyance.

10.6   From Noise Control to Perception-Driven Acoustic Engineering

Transportation noise has traditionally been managed using a noise control approach. This approach is mainly based on an assessment of decibels received at a receiver position, using a suite of noise metrics based on A-weighted equivalent sound pressure level integrated over a given time period ($L_{Aeq,t}$), or in some cases like sleep disturbance, event-based metrics like A-weighted maximum sound level (L_{Amax}). After an assessment has been done, appropriate (ad-hoc) interventions are designed and implemented to correct any exceedances of existing noise limits set by regulation. Such an approach usually provides a limited scope for solutions, as meeting a compliance level in dB does not consider the quality of the sound, and might not allow to address the core issue and meet communities' requirements and expectations.

The transition towards e-mobility, could offer policymakers and urban planners more scope for positive choices in the design of the urban sound environment. The acoustic design of the next generation of EVs and Advanced Air Mobility aircraft needs to incorporate not just models of sound emission and propagation, but also models of sound perception to understand how the sound will integrate into the overall soundscape. Embedding these models of sound perception into the design of novel vehicles can allow their optimisation to meet noise targets and psychoacoustic constraints at a conceptual level, and therefore avoid more costly and challenging ad-hoc solutions.

After being introduced by Davies and colleagues at Purdue University in 2007 (Davies 2007), several researchers and engineers have adopted a perception-driven engineering approach as a way to integrate human factors and perception into the design of engineered systems, and also have developed tools for its implementation to aid the design of vehicles and transport infrastructures. Examples of the transition towards perception-influenced engineering, or perception-driven engineering as proposed here, are the development (and consideration) of Sound Quality Metrics for a more holistic assessment of how sound is perceived (compared to A-weighted sound pressure levels) (Boucher et al. 2019); the development and implementation of psychoacoustic models (Fastl and Zwicker 2006; Torija et al. 2022); and the development of auralisation tools for the simulation of the noise produced by a given vehicle under expected operating conditions (Aumann et al. 2015). These auralisation tools have been suggested as a key element of perception-driven design of new aircraft (Rizzi and Sahai 2019); and road traffic (Finne 2016) and railway (Pieren et al. 2016) infrastructures.

This perception-driven approach has also proven to be useful for the design of alerting sounds for electric scooters (Walton et al. 2022), where a psychoacoustic model was used to optimise the design for maximum detectability and minimum annoyance. Further research and innovation for the continued development of perception-driven methods seem to be an inevitable requirement for shaping the future of mobility.

Therefore, if current methods are not optimised for better integrating human factors into the design of engineering systems and living spaces, the current energy transition will likely cause unintended effects in the form of decreasing human health and well-being due to new and unconventional noise sources.

References

Administration NHTS (2016) Minimum sound requirements for hybrid and electric vehicles: final environmental assessment (Document submitted to Docket Number NHTSA-2011–0100. Report No. DOT HS 812 347). National Highway Traffic Safety Administration (NHTSA), Washington, DC, 2016

Alexander WN et al. (2019) Predicting community noise of sUAS. In: 25th AIAA/CEAS aeroacoustics conference. Delft, The Netherlands

Aumann AR et al. (2015) The NASA Auralization framework and plugin architecture. 2015.

Berge T (2018) Experience and perception of AVAS on electric vehicles in Norway. In: Inter-noise and noise-con congress and conference proceedings. 2018. Institute of Noise Control Engineering

Boucher M et al. (2019) Sound quality metric indicators of rotorcraft noise annoyance using multi-level regression analysis. In: Proceedings of meetings on acoustics 177ASA. 2019. Acoustical Society of America

Cabell R, Grosveld F, McSwain R (2016) Measured noise from small unmanned aerial vehicles. In: Inter-noise and noise-con congress and conference proceedings. Institute of Noise Control Engineering

Campello-Vicente H et al. (2017) The effect of electric vehicles on urban noise maps. Appl Acoust 116:59–64

Christian AW, Cabell R (2017) Initial investigation into the psychoacoustic properties of small unmanned aerial system noise. In: 23rd AIAA/CEAS aeroacoustics conference

Clark C, Stansfeld SA (2007) The effect of transportation noise on health and cognitive development: a review of recent evidence. Int J Comparat Psychol 20(2)

Cotta J, Breque M (2021) Industry 5.0 - towards a sustainable, human-centric and resilient European industry. 2021, Directorate-General for Research and Innovation, European Commission

Davies P (2007) Perception-based engineering: Integrating human responses into product and system design. Bridge, Nat Acad Engin 37(3):18

DfT (2021) Reported road casualties Great Britain: e-Scooter factsheet year ending June 2021. UK Department for Transport

DfT (2022) Government response to the e-scooter trials evaluation report. UK Department for Transport

EEA, Decarbonising heating and cooling — a climate imperative. 2023, European Environment Agency.

Fastl H, Zwicker E (2006) Psychoacoustics: facts and models. Springer

Fiebig A (2020) Electric vehicles get alert signals to be heard by pedestrians: benefits and drawbacks. Acoust Today 16(4):20–28

Finne P (2016) Road noise auralisation for planning new roads. In: Inter-noise and noise-con congress and conference proceedings. Institute of Noise Control Engineering

Govindswamy K, Eisele G (2011) Sound character of electric vehicles. 2011, SAE technical paper

Gwak DY, Han D, Lee S (2020) Sound quality factors influencing annoyance from hovering UAV. J Sound Vibrat 115651

Hanna R (2009) Incidence of pedestrian and bicyclist crashes by hybrid electric passenger vehicles

Hansen C, Hansen K (2020) Recent advances in wind turbine noise research. In: Acoustics. 2020. MDPI

Hubbard HH (1991) Aeroacoustics of flight vehicles: theory and practice. volume 1. noise sources. 1991, National Aeronautics and Space Admin Langley Research Center Hampton VA

Iversen LM, Marbjerg G, Bendtsen H (2013) Noise from electric vehicles-'State of the art'literature survey. In: INTER-NOISE and NOISE-CON congress and conference proceedings. 2013. Institute of Noise Control Engineering

KANTAR, Perceptions of current and future e-scooter use in the UK: Summary report. 2021.

Karaaslan E et al. (2018) Modeling the effect of electric vehicle adoption on pedestrian traffic safety: an agent-based approach. Transp Res Part C: Emerg Technol 93:198–210

Laib F, Schmidt JA (2019) Acoustic vehicle alerting systems (AVAS) of electric cars and its possible influence on urban soundscape. 2019: Universitätsbibliothek der RWTH Aachen

Lennström D, Nykänen A (2015) Interior sound of today's electric cars: tonal content, levels and frequency distribution. 2015, SAE Technical Paper

Lennström D, Ågren A, Nykänen A (2011) Sound quality evaluation of electric cars: preferences and influence of the test environment. In: Proceedings of the Aachen acoustics colloquium

Lennström D, Lindbom T, Nykänen A (2013) Prominence of tones in electric vehicle interior noise. In: International congress and exposition on noise control engineering: 15/09/2013–18/09/2013. 2013. ÖAL Österreichischer Arbeitsring für Lärmbekämpfung

Liu X, Bo L, Veidt M (2012) Tonality evaluation of wind turbine noise by filter-segmentation. Measurement 45(4):711–718

Liu C, Zhao L, Lu C (2022) Exploration of the characteristics and trends of electric vehicle crashes: a case study in Norway. Eur Transp Res Rev 14(1):1–11

Madsen PT et al. (2006) Wind turbine underwater noise and marine mammals: implications of current knowledge and data needs. Mar Ecol Prog Ser 309:279–295

Marbjerg G (2013) Noise from electric vehicles–a literature survey. Report within Compett project, 2013

Miedema HM, Oudshoorn CG (2001) Annoyance from transportation noise: relationships with exposure metrics DNL and DENL and their confidence intervals. Environ Health Perspect 109(4):409–416

Mooney TA, Andersson MH, Stanley J (2020) Acoustic impacts of offshore wind energy on fishery resources. Oceanography 33(4):82–95

Muender M, Carbon C-C (2022) Howl, whirr, and whistle: the perception of electric powertrain noise and its importance for perceived quality in electrified vehicles. Appl Acoust 185:108412

Nguyen PD et al. (2021) Long-term quantification and characterisation of wind farm noise amplitude modulation. Measurement 182:109678

No, U.E.R., 138, Uniform provisions concerning the approval of Quiet Road Transport Vehicles with regard to their reduced audibility. Official Journal of the European Union, 2017. 13

Pallas M-A et al. (2016) Towards a model for electric vehicle noise emission in the European prediction method CNOSSOS-EU. Appl Acoust 113:89–101

Pedersen E, Persson Waye K (2004) Perception and annoyance due to wind turbine noise—a dose–response relationship. J Acoust Soc Am 116(6):3460–3470

Pedersen E, Waye KP (2007) Wind turbine noise, annoyance and self-reported health and well-being in different living environments. Occup Environ Med 64(7):480–486

Pieren R et al. (2016) Auralisation of railway noise: a concept for the emission synthesis of rolling and impact noise. In: Inter-noise and noise-con congress and conference proceedings. Institute of Noise Control Engineering

Rizzi SA, Sahai AK (2019) Auralization of air vehicle noise for community noise assessment. CEAS Aeronaut J 10(1):313–334

Rizzi SA et al. (2020) Urban air mobility noise: current practice, gaps, and recommendations

Schäffer B et al. (2021) Drone noise emission characteristics and noise effects on humans—a systematic review. Int J Environ Res Public Health 18(11):5940

Swart DJ, Bekker A, Bienert J (2016) The comparison and analysis of standard production electric vehicle drive-train noise. Int J Veh Noise Vib 12(3):260–276

Torija AJ, Clark C (2021) A psychoacoustic approach to building knowledge about human response to noise of unmanned aerial vehicles. Int J Environ Res Public Health 18(2):682

Torija AJ, Nicholls RK (2022) Investigation of metrics for assessing human response to drone noise. Int J Environ Res Public Health 19(6):3152

Torija AJ et al. (2019) On the assessment of subjective response to tonal content of contemporary aircraft noise. Appl Acoust 146:190–203

Torija AJ, Li Z, Self RH (2020) Effects of a hovering unmanned aerial vehicle on urban soundscapes perception. Transp Res Part D: Transp Environ 78:102195

Torija AJ, Li Z, Chaitanya P (2022) Psychoacoustic modelling of rotor noise. J Acoust Soc Am 151(3):1804–1815

Torija AJ, Li Z (2020) Metrics for assessing the perception of drone noise. In: e-Forum Acusticum 2020. 2020. Lyon, France: European Acoustics Association (EAA)

Torija AJS, Rod H, Lawrence Jack LT (2019) Psychoacoustic characterisation of a small fixed-pitch quadcopter. In: Inter-noise and noise-con congress and conference proceedings, InterNoise19. 2019. Madrid, Spain: Institute of Noise Control Engineering

Torija Martinez AJ et al. (2023) Generation and analysis of artificial warning sounds for electric scooters. In: Inter-noise and noise-con congress and conference proceedings. 2023. Institute of Noise Control Engineering

Verheijen E, Jabben J (2010) Effect of electric cars on traffic noise and safety

Wachter D (2009) Schallpegelmessungen an Elektrofahrzeugen („VLOTTE. Amt der Voralberger Landesregierung, Bregenz

Walton T, Torija AJ, Elliott AS (2022) Development of electric scooter alerting sounds using psychoacoustical metrics. Appl Acoust 201:109136

Waye KP, Rylander R (2001) The prevalence of annoyance and effects after long-term exposure to low-frequency noise. J Sound Vib 240(3):483–497

Wunderli JM et al. (2016) Intermittency ratio: a metric reflecting short-term temporal variations of transportation noise exposure. J Eposure Sci Environ Epidemiol 26(6):575–585

Yonemura M, Lee H, Sakamoto S (2021) Subjective evaluation on the annoyance of environmental noise containing low-frequency tonal components. Int J Environ Res Public Health 18(13):7127

Zajamšek B et al. (2016) Characterisation of wind farm infrasound and low-frequency noise. J Sound Vib 370:176–190

Chapter 11
How to Move Forward?

Irene van Kamp and Fred Woudenberg

A light rustling in the trees. The smell of eucalyptus. The whispering of pines… Space defined by sound; sound surrounded by silence. Spheres within spheres. There is no end there was no beginning. (Brink AP, States of emergency. London: Faber; 1988 (page 189).

Sound plays a key role in human life, in our orientation in time and space, a sense of safety or threat, survival, but also in the expression of emotion, such as in poetry and music and even in codetermining the meaning of life as in the Andre Brinks quote above. Our biological system is always active whilst awake and at sleep. So it is not surprising that chronic exposure to often mechanical and meaningless sound, that dominates our urban soundscape, can have a disrupting effect and in the long run can lead to physiological effects and diseases while pleasant sounds, like the sounds of nature in which humans evolved, support restoration and wellbeing.

Most books and reports about noise and health focus on the negative health effects of noise and only sometimes address the positive aspects. This book tries to look beyond this and include the different ways in which society can deal with unwanted and/or harmful sounds. In this chapter, we try to draw up the balance and describe what this book tells us about the best approaches to deal with sound and noise. Many countries, at least in Europe, have been working for decades now to improve the sound environment. This has certainly yielded positive results and decreased the exposure to unwanted and/or harmful sound. It has however not solved all problems nor led to the minimization of negative health effects that was hoped for at the time when countries first introduced their ambitious policies in the 70s aimed at the reduction of noise and annoyance. After more than 50 years of noise policies, it is still necessary

I. van Kamp (✉)
National Institute for Public Health and the Environment, Bilthoven, The Netherlands
e-mail: Irene.van.kamp@rivm.nl

F. Woudenberg
Municipal Health Service Amsterdam, Amsterdam, The Netherlands
e-mail: fwoudenberg@ggd.amsterdam.nl

© The Author(s) 2025
I. van Kamp and F. Woudenberg (eds.), *A Sound Approach to Noise and Health*,
Springer-AAS Acoustics Series, https://doi.org/10.1007/978-981-97-6121-0_11

to do research into the health effects of noise, to support its urgency and to devise and apply governance strategies to improve the urban soundscape.

11.1 Health Effects

What do we know about the health effects of sound? In Chap. 4 Charlotte Clark, Danielle Vienneau and Gunn Marit Aasvang give an overview. Worldwide billions of people experience severe annoyance and sleep disturbance due to noise related to road traffic, trains, neighbours, industry, aircrafts and other sources indoors as well as outdoors. Chronic exposure affects our wellbeing and disturbs our sleep leading to fatigue and reduced task performance in the daytime. Long-term this affects directly or indirectly the cardiovascular, metabolic and immune system causing cardiovascular diseases, obesity, diabetes type 2, increased risk of infectious diseases and cognitive effects in children.

The more research over larger populations is performed the broader the range of detrimental effects found. Several new issues have popped up in more recent years. For example, based on a literature review (Meng et al. 2022) an association was found between environmental noise and dementia. Also, there are strong indications that aircraft noise affects birth outcomes such as low birthweight, premature births as well as increases in breast cancer and evidence suggesting that high noise annoyance and poor mental health, in particular depressive symptoms, are interrelated. The direction of the association is not clear yet and it may very well be that poor mental health increases the response to noise and noise sensitivity. These effects are not only directly related to noise levels, but at the same time indirectly to the meaning of sound, covered by Rainer Guski in Chap. 3.

11.2 Meaning of Sound

Cars and airplanes produce anonymous, mechanical sounds, which almost everybody tends to dislike and qualify as noise. The perception of sound depends on much more than its acoustical properties alone. It is the meaning of a sound which determines whether it is annoying or enjoyable. Even with motor vehicles and airplanes, there are people who, at least in some situations, greatly enjoy their sounds. Plane spotters share videos of the best-sounding and loud takeoffs (https://www.youtube.com/watch?v=tPVGB8u6Z7w). Love of the characteristic sounds is a major constituent of Formula 1 enthusiasm, although this does not imply that it is not harmful. Even noise lovers can damage their ears.

The meaning of a sound is closely tied to the source of a sound. Intrusive and alarming sounds are of specific relevance to us, which is very understandable from an evolutionary perspective. The acoustics of sounds are tied to their meaning in subtle ways. Approaching (or "looming") sounds, i.e., sounds with increasing amplitude over a certain time are generally more alarming and annoying than so-called "receding" sounds, i.e. sounds with decreasing amplitude. If a person approaches us from behind the repetition rate of the sound of successive steps, together with the change in volume over time, provides information about the spatial proximity of the pursuer. The loudness, frequency spectrum, and onset and decay time of the friction sounds provide information about the person approaching us. The place, context and specific situation in which it happens, also gives meaning. Approaching footsteps heard in our living room may be welcome when heard as coming from outside at a time when we expect guests. The same sound can be alarming when we walk at night in a dark and neglected neighborhood with no other persons present.

A host of partly overlapping non-acoustic factors that are physical, personal, social, situational and contextual (e.g., infrastructural change situations) give meaning to the acoustic signals reaching our ears. About 10 to 15% of the population is highly sensitive to noise and as a consequence more easily annoyed by sounds. Fear or dislike of and distrust in the source of a sound are also important. A lack of personal control over a sound is linked to increased noise annoyance, diminished quality of life and increased health risks. Expectations about future increases in noise levels also have been shown to affect the level of annoyance. There are many other of these so-called non-acoustic factors interacting with each other and with the acoustic properties of sound to establish their meaning. The response to sound and noise is for a large part a learned response by repeated experience. This learning starts months before we are born and continues throughout our life.

Some people take the large influence of non-acoustic factors as evidence that noise annoyance is 'subjective' and that with some training you may come to like any sound. They believe that the perception of noise can disappear by changing your mindset. A wonderful example is that of the Dutch artist Sarah van Sonsbeeck who made an art project out of the noise annoyance caused by her neighbour upstairs (Sarah van Sonsbeeck 2010). She knew that the noise stopped if the neighbour was going to bed and dropped both shoes on the floor: thump … thump. One night she heard only one thump and she lay awake for a long time waiting for the second one before she could peacefully fall asleep. Noise annoyance caused by silence. It is not given to everyone and all the time to turn noise into art or something pleasant for another reason, not even for Sarah after finishing her project. The solution can only partly be found in the receiver. Something must be changed about the noise itself. Next to removing sources of noise as discussed above, the noise environment can also be made more pleasant in other ways.

11.3 Redistributing Sounds and Silences

As Marcel Cobussen points out in Chap. 9, three steps are needed to gain control over our sound reality; first, become aware of it, second, select those sounds which contribute to individual and societal well-being and third, combining sounds and silences in such a way that it contributes to our sense of belonging, feeling safe and pleasant. Although all of this seems to be an individual effort, many experiences, preferences and the relationships between sounds and their source are shared within groups of people and cultures. Despite the large individual, group and cultural differences, the appraisal of sound and noise, and their effects on people are quite similar. At the same time, we need to take geographical bias into account as most studies referred to in this book have been performed in Western countries and particularly in Europe. Typical features such as population density and access to green space and other geographical features differ between places and are not static. The same counts for the morphology of the built environment. Taken together they can strongly influence the restorative potential of a certain place.

11.4 Impact of Noise

The importance of noise for health can also be measured by comparing it to other (environmental) causes of disease. Chapter 5 on health impact assessment of noise by Juanita Haagsma and Mark Brink describes the summary measures of population health that make this comparison possible. Examples are the Healthy Adjusted Life Years (HALYs), Quality Adjusted Life Years (QALYs) and Disability Adjusted Life Years (DALYs). Of these, the DALY is used most widely. To be able to summarize all health effects of noise in one metric, it would be necessary to compare apples with oranges or maybe even chalk and cheese. Haagsma and Brink outline the problems and shortcomings of this in detail. An important one is that you must scale the different effects to a common metric. In the DALY this is the loss of 1 year of life. Other effects get a weight between 0 and 1 corresponding to the severity of the disease (sometimes weights higher than 1 are used, if living for a year in an extremely severe and painful condition is evaluated as worse than dying). The main health effects contributing to the DALY score for noise are severe annoyance and sleep disturbance. The weights for these are set by experts or a lay audience. For noise annoyance the weight is often set at 0.01 implying that being severely annoyed by noise for 100 years is equivalent to dying 1 year earlier. Setting the weight at 0.02 results in a doubling of the DALY contribution of severe annoyance for noise. Some people have their doubts about the validity of making such comparisons. The role of annoyance and sleep disturbance is controversial since these outcomes are not based on an International Classification of Diseases (ICD) code (Van Kamp et al. 2018). Be that as it may, the DALY offers a unique opportunity to score the 'seriousness' of noise as a health effect and to compare it to other determinants of health. It can also be used to rank the sources of noise. Several such comparisons have been made internationally and nationally. Results over large populations show that road traffic contributes most to noise DALYs. DALY calculation are often used in the priority setting of policy measures aimed at population health gain. Comparisons that have been made show that in the environmental realm noise ranks second in terms of health impact after air pollution. In the list of all determinants of health, noise scores roughly at the same level as the use of alcohol, but after main determinants like smoking, unhealthy diet and lack of exercise (Integratiematen voor de Volksgezondheid Toekomst Verkenning (VTV) 2018).

The DALY approach is highly related to measures calculating the cost and benefits of environmental noise or interventions aimed at reducing the effects, as addressed by Ronny Klæboe in Chap. 6 on economics. Several economic valuation methods exist that can be applied to noise. The simplest ones are cost-effectiveness analyses that can be used to allocate resources and prioritize measures to decrease noise levels. These are especially suitable if there is no discussion that a measure must be taken, for instance when it is mandatory, and the search is for the most effective intervention. In most situations, noise is just one of many aspects, often not even the most important one, and choices can and must be made. A cost–benefit analysis then must be done. Such an analysis must involve all costs and benefits of noise-decreasing measures. For this to work, the cost–benefit analysis should be used in urban planning and ideally must be done at the beginning of a planning process, informing the decisions to be made. Hereby noise must be weighed against other aspects of the process and there should be tight cooperation between noise experts and the many other experts and stakeholders involved. Outcomes of economic valuation methods can also be used for devising taxation schemes. A tax could then be set at a sum equal to the marginal damage costs that the activity causes, the so-called Pigouvian tax. These costs thereby become internalized and are no longer neglected.

11.5 Noise an Underestimated Determinant of Ill Health?

It may seem surprising that in a world where people attach great importance to health, often defining it as the greatest good, there is still massive exposure to noise, despite decreases in some sources. Klæboe provides us with a good example of Norway in Chap. 6. If sound is so basic to health and noise has such a great negative impact on health, why are our cities still filled with noise instead of pleasant sounds. In most European countries the environment has improved in many ways. The swimming water quality has improved significantly, the air quality has become much better. Air quality is the most interesting to compare with noise, since they share the same main sources. Road traffic is the main source of air pollution and noise everywhere and the main contributor to the DALY score. Figure 11.1 shows how levels of particulate matter have been decreasing since 2005. This is an extension of a much longer trend that can be observed in many European countries since the nineteen seventies. Why have levels of air pollution dropped so significantly while overall the noise levels remained the same or increased (Noise pollution and health 2023)?

Society attaches more importance to air pollution than to noise pollution. Air pollution is at the center of attention of the public, media, scientists, (public) health professionals, politicians, and regulators whilst noise gets its share of attention, but much less. One explanation could be that pollution from cars can be reduced relatively easily by measures at the source. European regulations step by step decreased the allowable emissions of cars, from the Euro 1 to at present Euro 7 standards and

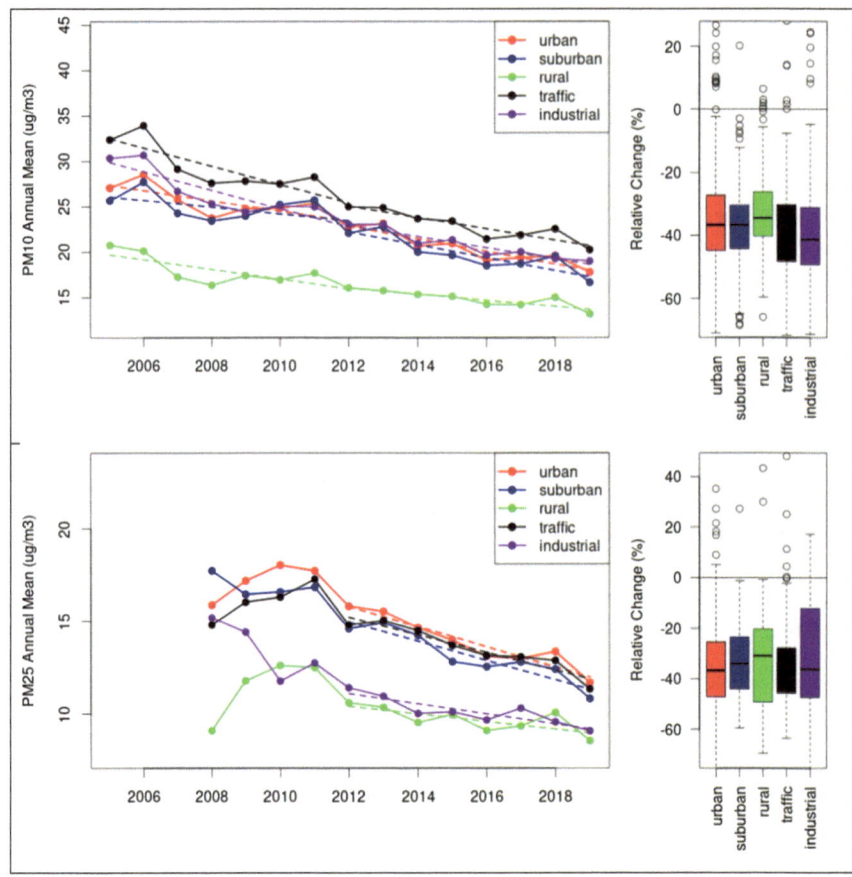

Fig. 11.1 Long-term trends of air pollutants at the national level in the period 2005–2019. (Solberg et al. 2021)

these regulations have been implemented successfully. For noise, the challenge is bigger. While fossil fuel vehicles with loud engines are being replaced by electric vehicles with quiet engines, the tires still make noise and in some cases even more due to the higher weight of the car. There is no simple technical solution for this. Noise measures must be taken farther away from the source with for instance noise barriers, zoning and insulating houses. These measures are often costly, occupy valuable city space or are technically complicated. In the case of noise, the source cannot always be silenced easily.

11.6 Is Annoyance Subjective?

People often dislike noise because they dislike the causative agents of it. Examples abound to illustrate this (Maris et al. 2007) and Maag and Gisladottir mention a few in their Chap. 7. The noise produced by people you know very well and who you like very much are seldom annoying. This often goes along with the possibility to influence the sounds. A study in The Netherlands (Devilee and Kamp 2013) showed mopeds to rank highest in the sources of noise causing annoyance. A reason given was that the high-pitched sound of the moped accelerating or stopping quickly signals danger. In both situations, a focus on lowering the noise level can be helpful, but hard to achieve. Removing the source can certainly be helpful, but is even harder to achieve. Getting rid of antisocial neighbours is extremely difficult. Electrification of mopeds and other transportation modes will certainly help, but this takes a lot of effort. Improving the relationship with your neighbours and the youngsters in your neighborhood is probably the most feasible way to solve your problems and therefore this is often applied. In many places, local governments and foundations offer mediation and organize activities to improve relationships. This may result in lower noise levels, but not necessarily. Improving the relationship with your immediate neighbours and the people living in your neighborhood, will lead to less noise annoyance even if the noise levels are unchanged and is good for social cohesion as well.

11.7 How Does the Future Sound?

Our urban soundscape is changing and will be characterised by new sources and their characteristics in the near future. Our current estimates of impact as described by WHO (World Health Organization 2018) and more recent updates are already outdated and might not apply to these new sources as we can observe now for example in relation to the health effects of wind turbine-related noise. We know that already at lower noise levels people are annoyed by wind turbines and the amplitude modulation might be an important determinant, rather than the often suggested effect of the lower frequencies. In Chap. 10 Antonio Torija Martinez describes how future changes in road and air traffic may alter the urban soundscape unrecognisably.

Future road traffic will be dominated by electric cars. An electric car has no tailpipe and therefore no exhaust gases. Electric cars still cause some air pollution because of tire shavings producing what are called Tyre and Road Wear Particles (TRWP) mainly in the PM2.5 part of particulate matter. These tires of the relatively heavy electric vehicles still produce a lot of noise. Because electric motors hardly produce any noise, many people expect (and hope) that the large scale introduction of electric vehicles will silence the cities. Antonio Martinez shows in Chap. 10 that this hope is in vain: little reduction in noise from the transition to electric vehicles is to be expected. Noise annoyance could even increase. Electric vehicles differ from fossil fuel vehicles in noise spectrum and character, with electric vehicles having more high frequencies which people find more annoying. Also, the quieter e-drive can reveal other, especially tonal sounds which were previously masked. For safety reasons, especially for the blind, sounds must be added to limit the number of collisions, so-called Acoustic Vehicle Alerting Systems (AVAS). Each type of vehicle (truck, car, scooter) must have its own tune to be able to distinguish them. According to Martinez, this can lead to dissonant and inharmonious urban soundscapes which could increase noise annoyance. And this is the case even without considering the possible development of the classic science fiction prediction of flying cars. Within decades the electric Vertical Take-off and Landing (eVTOL) vehicles may become fact instead of fiction. Martinez explains that they have more tonal, high frequency and time-varying sounds which are unpleasant to most people. If the large-scale introduction of drones with similar sound characteristics is added to the future urban soundscape, firm measures are certainly needed to prevent an increase in noise instead of the hoped for decrease with the coming electrification of land and air traffic.

11.8   Governance and Planning

Technological developments in the past and future alone will not dramatically improve the urban soundscape. Noise has been and will be an important cause of negative health effects when mitigating measures are not taken. Substantial improvement can only be attained by firm measures through governance and urban planning, the topic of Chap. 8 by Benjamin Fenech and Nathalie Riedel.

Important national and international political actors coming forward are organizations, such as the UN (Environment Programme; International Civil Aviation Organization; Economic Commission for Europe), OECD and WHO. They have published reports with recommendations and suggestions for policies to abate noise or to create positive sound environments. WHO developed noise guideline levels for specific health effects and for specific environments, as was done for air quality, but with a method which is even more rigid and thorough. The United Nations International Civil Aviation Organization (ICAO) and Economic Commission for Europe (UNECE) also have set standards for the sound level of airplanes and vehicles, respectively. According to the latest revision with the introduction of a new test procedure which more realistically reflects the exterior noise of vehicles in typical urban traffic situations, the sound energy is said to have dropped by 10 dB since the regulation originally came into force in 1982 (https://unece.org/press/unece-world-forum-harmonization-vehicle-regulations-tightens-vehicles-noise-limits-and-adopts). However, there are two downsides: there is more noise at traffic lights and intersections and the effect of quiet road surfaces wears off in time due to a lack of maintenance. Calculation methods should take this into account. Otherwise, they lead to underestimation of noise levels.

International organizations for standardization like ISO (International Organization for Standardization) and IEC (International Electrotechnical Commission) set standards, sometimes modified by (supra)national bodies. They do this for amongst others the specification and calibration of equipment, measurement, and research methods (for sound, soundscape, annoyance etc.) and the acoustic classification of dwellings.

The above are all activities by 'nonpolitical' organizations. The European Union has the European Noise Directive (END), which obliges member states to make noise action plans and noise maps, but it does not set source-specific limit values. This is delegated to the member states themselves.

The bulk of the activities described above are on procedure and methodology. Applying these to really limit sound levels or optimize urban soundscapes must come from government action at local, regional, and national levels. Two important lessons can be learned from Chap. 8 by Fenech and Riedel. One is that limit values are rarely set, even in Europe, or are very lenient and thus providing room for many exceptions. The other is that noise governance often is part of broader policies concerning economic development, urban planning, mobility etcetera. Many European countries have some form of noise regulations including noise mitigation measures such as insulation of dwellings. The siting of industrial plants and wind parks is in general more controlled than road traffic. The broad picture is that all European countries have noise regulations, but they are not extremely strict and mostly subordinate to other policy domains. Regulations being enforced can be observed in Australia, New Zealand, and some parts of Asia. The interesting case study Fenech and Riedel give for Nepal indicates that the situation can be worse in other regions.

11.9 Integrating Soundscape into Urban Planning

Urban planning is an important domain where sound comes into play. Currently, most people in the world live in a city. This means that cities over the world face a multitude of challenges, and the sound environment is one of them. The way the growing number of inhabitants transport themselves is a major driver of the urban soundscape.

Trond Maag and Arnthrudur Gisladottir in Chap. 7 on urban planning see two ways in which cities can develop: the compact or the dispersed city. The compact city is extremely lively and in combination with the dense built-up can be very noisy, but at the same time, buildings can function as a shield to roads, industries and other noise-generating activities. In the dispersed city there is much open (green) land between built-up areas in which sound can sprawl over long distances and affect a wider area than in denser cities.

Bringing down urban noise exposure levels is a significant challenge, and enormous efforts are needed, especially if noise is not or is only considered late in the planning process. Noise regulations are predominantly targeted at creating quiet facades below a certain noise limit. Reducing traffic, lowering traffic speed, changing traffic patterns, applying low-noise road surfaces and shielding can be effective to achieve this, but are difficult to realize. Other possible measures are zoning and optimizing floor plans in buildings. As all of this quickly hits its limits, planners seek the solution in the measure that is applied most often: facade insulation and where space is scarce (as often is the case in growing cities) in granting exemptions.

When noise is made a major consideration in urban planning, substantial improvements in the sound of the city can be achieved. The soundscape approach offers huge opportunities to make cities sound pleasant, but to put this into practice, planners and developers need soundscape expertise, guidance on identifying scenarios for soundscape actions, and clarification on how to identify specific objectives of a soundscape design at a specific place and translate them into design criteria. Quiet green public spaces, including small 'pocket parks', can be important positive elements in the urban soundscape of a neighborhood, but also not always easy to realize. Green spaces have a lot of beneficial effects apart from offering quiet. They also influence the perception of noise, as people perceive roads and railways as up to 10 dB(A) less noisy the greener the neighborhood (Klæboe et al. 2004).

Invoking noise reduction objectives and soundscape approaches in regular urban design and planning practices offers the largest opportunity to create pleasant-sounding cities with minimal noise annoyance. Knowledge about sound must be made accessible for planners. This is not only about the sound levels of cars and other sources. The influence of social relations and human behavior on the production and perception of sound must be considered as well, making it necessary to involve local communities.

Considering sound and noise must be included in all levels of urban planning: from the master plan for large areas, to design briefs at the neighborhood scale and finally the design of buildings, roads, etc. Personal commitment of the planners in charge helps. Maag and Gisladottir mention four examples where the local community was involved resulting in better plans, happier users, and increased trust in government. Although these are still exceptions, much is to gain by involving communities and taking sounds as an important determinant of city quality.

11.10   An Integrated Design Approach

Maag and Gisladottir make clear that sound is always a part, and often only a small one, of a broader picture in the goings and development of cities. A limited focus on sound alone then is not the most effective approach to combat noise. Measures for threats to public health need to reflect interactions with the wider determinants of health, including potential co-benefits and unintended consequences of regulations and interventions. Also, it should be accounted for that sound and noise interact with the domains of lifestyle, community, local economy, activities, built environment, and natural environment for which a holistic approach is needed. The many overlaps of public space present an acoustic challenge for its design and organisation, which at the same time must also meet environmental objectives in terms of city climate, urban ecology and biodiversity, and social objectives. The disciplines intersecting in public space call for more integrated design approaches, rather than separate ones.

Since the increase in traffic volume and the changes in vehicle size and power have been major reasons why noise annoyance and sleep disturbance have not declined, a reduction in vehicle traffic is a highly effective way to improve the soundscape of urban and other environments. There are ample arguments for curtailing traffic of which noise is an important, but certainly not the only one. An important reason is air pollution (although this will not disappear with the electrification of traffic because of tire wear particulate matter). Cars are also the main cause of traffic accidents. Fewer cars probably have a larger effect on safety than equipping them with alerting signals.

Parked as well as on the road, cars also occupy a lot of space, as is illustrated in Fig. 11.2. Space in growing cities is extremely limited, especially since many cities choose the compact alternative of the two possible ways in which cities can develop-sprawl versus compact cities. Consider, for instance, the 15-min city goal of Paris with a large density of houses and facilities (including green space). Often there is simply no room for more cars as adding more leads to extreme congestion.

As part of their effort to improve the health of their inhabitants, many cities are stimulating active transport: walking and bicycling. This can only be accomplished when accompanied by a reduction in the least active transport mode: the car. Sustainability is also a good reason to limit car use. Electric cars running on green energy still use raw materials to be produced. Green energy is limited and must be used efficiently and driving a car is not the most efficient way to use it.

Fig. 11.2 In 2012 69 volunteers, 69 bicycles, 60 cars, and one bus gathered in Canberra, Australia for this world-renowned photograph to demonstrate the advantages of bus and bicycle travel in congested cities. Photo by Andrew Taylor, courtesy of we ride Australia (https://www.weride.org.au/events/the-power-of-an-image-the-canberra-transport-photo/)

Taking this all together it is no surprise that many cities in the world limit car traffic and expand their car-free surface (Kersley 2022). Noise is never the main reason to do so, but less noise is almost always the result. With air traffic, noise is an important reason to want to limit it, but certainly not the only one. The emission of CO_2 and its large contribution to global warming is the main concern here. Airplanes also cause air pollution mainly in the form of ultrafine particles in the vicinity of airports.

A sound approach to noise and health is more than an approach that focuses solely on sound and noise. There is no simple technical solution to silence sources of noise. There are no silent cars and silencing them takes a lot of effort. Curbing noise annoyance by cars costs money and valuable space for zoning or larger housing adds volume and speed. For most decision makers sound and noise are simply not important enough to make these sacrifices. Money and space are scarce and in the compact city space claims are fighting for priority constantly.

At the same time, sound and noise must be considered as important. Existing regulations to limit sound levels achieved some improvements. Measures that have been taken to lower the sound levels of sources (silent airplanes for instance), reduce the transmission of sound (noise barriers) and the reception of sound (zoning, insulation of houses) have been highly effective. These measures prevented the increase in volume of noise-producing machinery and activities (traffic, leisure, city density, etc.) accompanied by an equivalent rise in the percentage of people experiencing the negative health effects of noise. The measures only curbed the increase but did not reduce the overall number of people affected in terms of annoyance, sleep, and long-term health effects. To achieve this, additional approaches limiting the volume of noise and enhancing health-promoting sounds are needed. This can only be done in a governance climate and with an urban planning process in which sound is not the only, but an important factor to consider, together with other health-promoting factors like green, active transport, and recreation. In the plans of local and national governments, the urban environment is always pictured as lush green with friendly people walking, bicycling, or relaxing. There are no cars running over pedestrians, racing mopeds, lowly overflying airplanes (or VTOLs) and the sky isn't clouded by drones. To really improve the sound environment, the lush green vistas must be put into reality.

11.11   Conclusion

Summing up the state of the art we see an overall increase in noise exposure, but at the same time, an increasing awareness that noise directly or indirectly via annoyance, stress, and sleep disturbance have long-term health effects. Combinations of often small measures can lead to large reductions in noise levels and their negative health impacts. The long-term cost of health outweighs the cost of these interventions.

Efforts of people working in the world of noise and sound to make these reductions happen remain necessary. Many of them have been working for many years tirelessly to get noise higher on the public agenda. A small number of them have contributed to this book. We hope our 'Sound Approach to Noise and Health' will help to create a future in which fewer people suffer the negative effects of noise and more people enjoy the sounds that restore their well-being in pleasant soundscapes sometimes surrounded by silence!

References

Devilee JL, van Kamp I (2013) Noise annoyance caused by mopeds. About decibels and meanings. briefrapport 630650006/2013 (in Dutch). RIVM, Bilthoven, The Netherlands

https://www.weride.org.au/events/the-power-of-an-image-the-canberra-transport-photo/

https://www.youtube.com/watch?v=tPVGB8u6Z7w

Integratiematen voor de Volksgezondheid Toekomst Verkenning (VTV) (2018) Resultaten en methodologie. RIVM, Bilthoven, The Netherlands (in Dutch)

Kersley A (2022) People hate the idea of car-free cities—until they live in one. Wired. https://www.wired.co.uk/article/car-free-cities-opposition

Klæboe R, Amundsen AH, Fyhri A, Solberg S (2004) Road traffic noise–the relationship between noise exposure and noise annoyance in Norway. Appl Acoust 65(9):893–912

Maris E, Stallen PJ, Vermunt R, Steensma H (2007) Evaluating noise in social context: the effect of procedural unfairness on noise annoyance judgments. J Acoust Soc Am 122:3483–3494. https://doi.org/10.1121/1.2799901

Meng L, Zhang Y, Zhang S, Jiang F, Sha L, Lan Y, Huang L (2022) Chronic noise exposure and risk of dementia: a systematic review and dose-response meta-analysis. Front Public Health 10:832881. https://doi.org/10.3389/fpubh.2022.832881. eCollection 2022

Noise pollution and health (2023) EEA https://www.eea.europa.eu/publications/zero-pollution/health/noise-pollution

Solberg S, Colette A, Raux B, Walker SE, Guerreiro C (2021) Long-term trends of air pollutants at national level 2005–2019, ETC/ATNI Eionet report 9/2021. European topic centre on air pollution and climate change mitigation

https://unece.org/press/unece-world-forum-harmonization-vehicle-regulations-tightens-vehicles-noise-limits-and-adopts

Van Kamp I, et al (2018) Study on methodology to perform environmental noise and health assessment. In: RIVM report 2018–0121. Bilthoven, The Netherlands

Van Sonsbeeck S (2010) Bakery, 23 October-23 December 2010-Overview. Annet Gelink Gallery

World Health Organization (2018) Environmental noise guidelines for the European region. WHO Regional Office for Europe, Copenhagen, Denmark. https://www.who.int/europe/publications/i/item/9789289053563